Günther Gademann

NMR-Tomography of the Normal Brain

With 68 Figures

Springer-Verlag
Berlin Heidelberg New York Tokyo 1984

Dr. med. Dipl. Phys. GÜNTHER GADEMANN
Universitäts-Strahlenklinik
Voßstraße 3
D-6900 Heidelberg 1

ISBN-13: 978-3-642-69622-0 e-ISBN-13: 978-3-642-69620-6
DOI: 10.107/78-3-642-69620-6

Library of Congress Cataloging in Publication Data. Gademann, Günther, 1952– NMR-tomography of the normal brain. Bibliography: p. Includes index. 1. Brain-Anatomy-Atlases. 2. Brain-Radiography. 3. Nuclear magnetic resonance-Diagnostic use. 4. Tomography. I. Title. [DNLM: 1. Nuclear magnetic resonance-Methods. 2. Tomography, Emission computed. 3. Brain-Radionuclide imaging. WL 300 G124n] QM455.G24 1984 611'.81'022 84-5417

Softcover reprint of the hardcover 1st edition 1984

Printing and bookbinding: Appl, Wemding
2121/3130-543210

To my wife Gisela and
to my children Katharina and Georg

Preface

This book is intended as a short guide to the visualization of the anatomy of the normal brain by means of the NMR tomogram. The first section comprises a brief introduction to the physical and technical aspects of NMR. This is followed by the atlas section, which pursues a number of objectives. On the basis of two important NMR imaging techniques, the spin-echo technique and the inversion-recovery technique, those experienced in CT are given the opportunity to familiarize themselves with the differences in tissue contrast that exist in NMR scans despite their apparent similarity to conventional CT scans. The mode of action of the two NMR imaging techniques is explained in the technical introduction. An additional innovation is the possibility of producing sections that are not, as in CT scanning, limited by the body of the patient. The sagittal and frontal sections parallel to the plane of the face can show an unfamiliar, but particularly clear, image of the anatomy of the head and brain compared with conventional horizontal sections. An anatomical description accompanying every section is provided by way of clarification.
A particular advantage of NMR imaging, namely, the absence of ionizing radiation and, thus, an injurious effect on biological systems (Budinger 1981), makes it possible to provide a systematic visualization of a healthy human brain in a living person in three planes, arranged at right angles to one another. The use of different projections permits even complicated anatomical structures to be accurately defined and inspected. Since bone generates only a very low NMR signal, the NMR tomogram produces an astonishingly clear image of the posterior cranial fossa without the bony artefacts that plague the CT scan.
Contraindications for NMR imaging, such as claustrophobia, cardiac pacemakers (Pavlicek et al. 1983) or implanted ferromagnetic substances (New et al. 1983), did not apply to our examinations and will probably be of only secondary importance in the clinical setting.
The excellent contrast of the images, which were all obtained without the use of contrast medium, makes it possible to identify the anatomical structures in the noncontrastive image, which represents the picture as seen by the observer of the monitor screen. Importance was also attached to a differentiation of the structures of the face and neck in an attempt to address not only neurologists and neurosurgeons but also ENT specialists and oral surgeons, who will certaintly see in NMR imaging an interesting addition to their diagnostic possibilities.
I should like to take this opportunity to express my sincere thanks to the Medical Engineering Group of Siemens at Erlangen for kindly making their 0.2 Tesla machine available to me and for providing much needed support on questions pertaining to physics and engineering, which was

readily given by those working in the relevant areas. I am most grateful to Christiane, whose willingness to play the guinea pig made these fine scans possible. Finally, I would like to express my warmest thanks to Prof. Fahlbusch and Prof. Huk of the Neurosurgical Hospital of the University of Erlangen, West Germany and Prof. Rohen of the Anatomical Institute of the University of Erlangen for their support and for critically reading the manuscript.

Heidelberg, Spring 1984 GÜNTHER GADEMANN

Contents

1. Physical and Technical Fundamentals

Physical Aspects of Nuclear Magnetic Resonance

As early as 1946 the two physicists Bloch and Purcell described, independently of each other, the nuclear induction experiment (Bloch 1946; Bloch et al. 1946; Purcell et al. 1946). In the subsequent years, in the form of nuclear magnetic resonance (NMR) spectroscopy, it developed into one of the most important techniques employed in chemistry for the clarification of complicated chemical compounds. Initial ideas on the possibility of utilizing NMR for imaging purposes, proposed by Damadian (1971) and Lauterbur (1973) at the beginning of the 1970s, were followed by some 10 years of intensive developmental work, the result of which has been to make available to modern medicine a new high-quality diagnostic technique. The sectional images of the human body obtained on the basis of the principles involved are characterized by excellent contrast, which is achieved without the need for contrast medium, and without any of the damaging side effects typical of ionizing radiation.

This procedure makes use of a particular property of numerous atomic nuclei, namely, their inherent rotation, also known as spin. This is a property of all nuclei that have an odd number of nuclear particles (protons and neutrons), and it enables them to be utilized in NMR experiments. In accordance with the Maxwell equations, a moving charge generates a magnetic field, so that the rotating, positively charged atomic nuclei also possess a magnetic moment, μ. It is, however, not possible to measure any "macroscopic" field over a sample, since the tiny magnets are, statistically, arranged in all spatial directions, so that the "microfields" cancel one another out (Fig. 1.1).

A stationary magnetic field, B_o, imposed on the object to be measured results in an alignment of the magnetic (nuclear spin) axis in the direction of the external field. The result is now a measurable magnetic moment, the size of which increases with the strength of the superimposed magnetic field and decreases with increasing temperature of the sample (Fig. 1.2). An applied radio frequency (RF) field of a given frequency, the polarization direction of which is perpendicular to B_o, causes the aligned spins (magnetic moments) to precess (move like small gyros) around the axis of the static magnetic field (Fig. 1.3). The resonant frequency required for this – also called the Larmor frequency, ν_L, of the system – can with the aid of quantum-mechanical computations be determined in accordance with the equation:

$$\nu_L = \gamma \cdot B_o/2\pi \qquad (1.1)$$

Thus, the resonant frequency increases in linear fashion with the applied static magnetic field and a nuclear-specific constant, γ, the gyro-magnetic

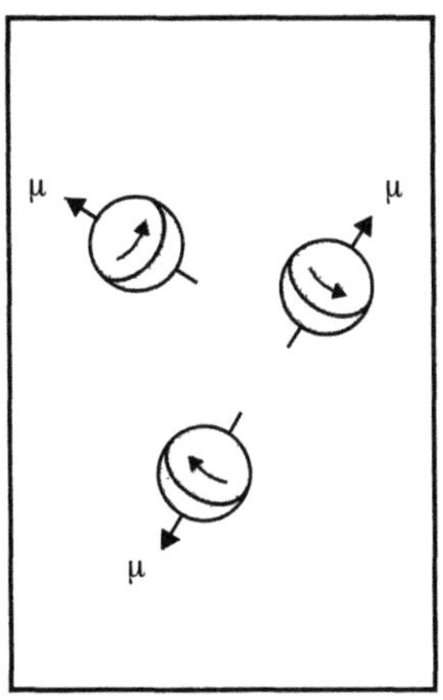

Fig. 1.1

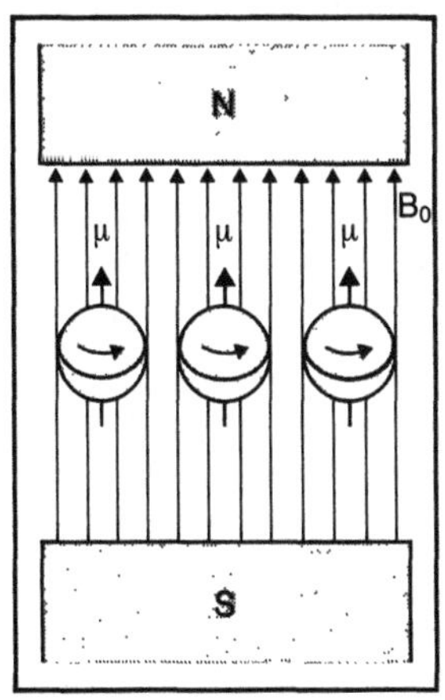

Fig. 1.2

Fig. 1.1. Disordered motion of the spins (spin directions) and thus of the magnetic moments in the absence of an external magnetic field. (After Habermehl and Graul 1982)

Fig. 1.2. Alignment of the spin axes along the field lines of the magnetic field B_o. (After Habermehl and Graul 1982)

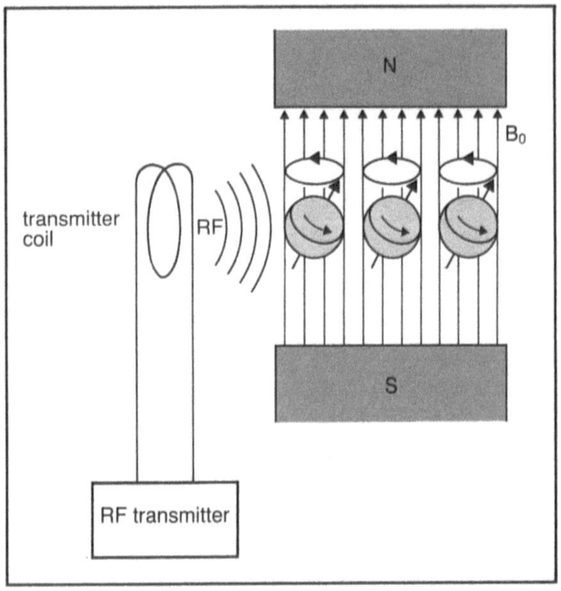

Fig. 1.3. Excitation of the precessional motion by the application of an RF field with the aid of a transmitter. (After Habermehl and Graul 1982)

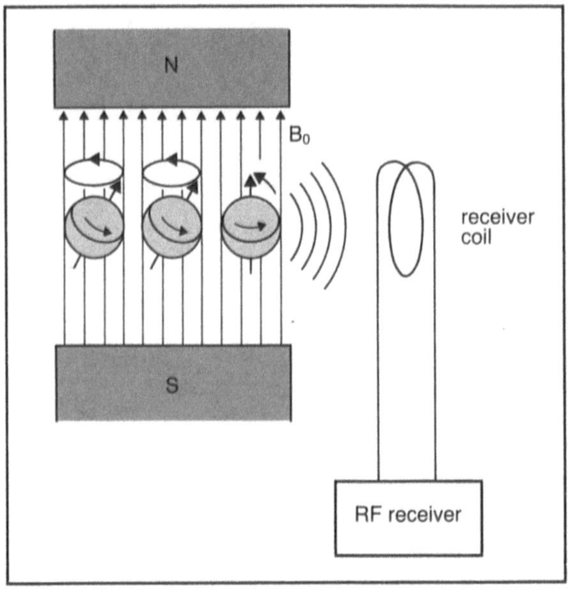

Fig. 1.4. Emission of a radio frequency signal during relaxation, and induction of a voltage in the receiver coil. (After Habermehl and Graul 1982)

ratio. Equation (1.1) is called the resonance condition of the NMR experiment.

When the excitation (RF) pulse is switched off the precessing atomic nuclei emit electromagnetic radiation, the frequency of which is identical with the Larmor frequency. The amplitude of the emitted high-frequency field decreases with a certain time constant. During this so-called relaxation the resonant signal, which is used to produce the NMR image, is measured (Fig. 1.4).

The decrease in the signal amplitude, equivalent to the temporal behaviour of the spin axes realigning themselves to the axis of the static field, has an exponential course and is influenced by two mechanisms. The spin lattice or longitudinal relaxation is brought about by interactions of the nuclear spins with the surrounding molecular structures and forces the precessing nuclei to return to their thermal state of equilibrium. The spin-spin or transverse relaxation comes about as a result of the mutual interactions of the nuclear spins; it results in a gradual "falling out of phase" of the "gyros", which at the end of excitation all precess uniformly, and in this way also brings about an attenuation of the signal amplitude. Since this requires a precession of the nuclei the transverse relaxation is always associated with a shorter time constant, T_2, than is the longitudinal relaxation with its time constant T_1. These relaxation times show a tendency to increase with the size of the static field.

The hydrogen nucleus is the most suitable for NMR imaging because of the high gyro-magnetic ratio of 42.5659 MHz/T of the proton and because of its prevalence in organic substances. Phosphorus 31, which on account of its high-energy compounds that are involved in the metabolism of the cell would be of considerable interest, requires a static field of more than twice the strength of that used with the H nucleus to acheive the same resonant frequency. Furthermore, it is present in the organs of the body at considerably lower concentrations. As a result, phosphorus 31 imaging produces only a poor spatial resolution and a low signal-to-noise ratio (Bendel et al. 1980). In the case of the hydrogen nucleus, at a field strength of 0.2 T the longitudinal (T_1) relaxation times typical of biological tissue are between 200 and 1500 ms, the T_2 times being between 20 and 500 ms (Huk et al. 1983; Rupp et al. 1983).

The resonant frequency is measured within a given volume. The size of the measured signal depends upon the number of atomic nuclei within the measured volume that can be induced to produce a resonant signal, that is, the nuclei (magnetic moments) aligned in the direction of the static field axis. This means that the measured signal depends on the size of the measured volume and of the magnetic field, B_o. An image that is made up of numerous small image points (pixels) requires a maximum of tiny volume elements in order to achieve adequate sharpness. However, the diminishment of such a picture element is limited by the signal-to-noise ratio, which in turn is determined by the unavoidable thermal and electronic noise. The signal can be enhanced by increasing the strength of the magnetic field. In accordance with the resonance equation (Eq. 1.1) this also results in an increase in the resonant (Larmor) frequency. In material such as that of the human body electromagnetic alternating high-frequency fields undergo both attenuation and a phase shift, which eliminate the positive effect of increasing the strength of the field (Bottomley and Andrew 1978). At the present time the frequencies commonly employed for proton tomography are between 5 and 20 MHz, that is, radio frequencies in the short-wave band. Such frequencies have a wavelength of several metres, so that a direct visualization of the organs within the body is impossible. Shaped magnetic fields have to be employed to achieve a spatial encoding of the measured signals.

With the aid of NMR imaging, therefore, not only the density of the excited atomic nuclei but also properties associated with the chemical structure are measured. The most important contrast-generating magnitudes are the already-mentioned relaxation times, T_1 and T_2. They provide formation as to the physicochemical environment of the hydrogen ions. Thus, for example, they are determined to a large extent by the viscosity of the surrounding milieu, the content of paramagnetic salts and the type of bonding of the protons.

Another method is spectroscopy, in which the frequency spectrum of the emitted signal is investigated. Frequency shifts (chemical shifts) provide considerably better information about the chemical fine structure than do the relaxation times of NMR imaging. In this connection, two nuclei are of particular interest, the already-mentioned phosphorus 31 (Ng et al. 1982; Evanochko et al. 1983) and the stable carbon isotope, ^{13}C, which accounts for about 1% of naturally occurring carbon. These two atoms are involved in numerous biochemical processes, so that their study can be expected to lead to direct and comprehensive insights into metabolic mechanisms. Since for the resolution of the spectral lines both higher frequencies and larger and more homogeneous magnetic fields are required, the technology employed must be more complicated than that required for proton tomography. The application of NMR spectroscopy in medicine is still in its infancy, and any discussion of its possible value is beyond the scope of this book.

For the first time, the diagnostic imaging technique of NMR tomography permits a direct insight into the chemistry of living matter without, as far as is known at present, producing any injurious effects in the organism (Budinger 1981), and it provides information about an important organic building block, the hydrogen atom. In contrast, the important parameters in X-ray computerized tomography (CT), namely, the mass density of the tissue, or the different impedances of various tissues that are utilized in ultrasound examinations appear coarse and unspecific.

Technical Features of NMR Imaging Equipment

The major component of a modern NMR imager is the large, current-carrying coil which generates the static magnetic field, B_o (Fig. 1.5). At present, field strengths of about 0.1–0.5 T – this is about ten thousand times the strength of the earth's magnetic field (0.06 mT) – are usually employed. For the generation of a field strength of up to 0.2 T a normally conducting copper coil suffices; the power lost in the form of heat, and amounting to several kilowatts, has to be dissipated via a cooling system. For greater field strengths supraconducting magnets are required. In such a case, although very little electric power is consumed the entire coil has to be cooled to very low temperatures (4.2 K). Accordingly, the technology involved is highly sophisticated, but it has the advantage of producing a stronger resonant frequency signal and the possibility of utilizing high-resolution spectroscopy. The required highly homogeneous magnetic field is generated within the centre of the coil and determines the measuring volume. The patient lies in the direction of the longitudinal axis of the coil, that is, parallel to the z direction of the field, as shown in Fig. 1.5.

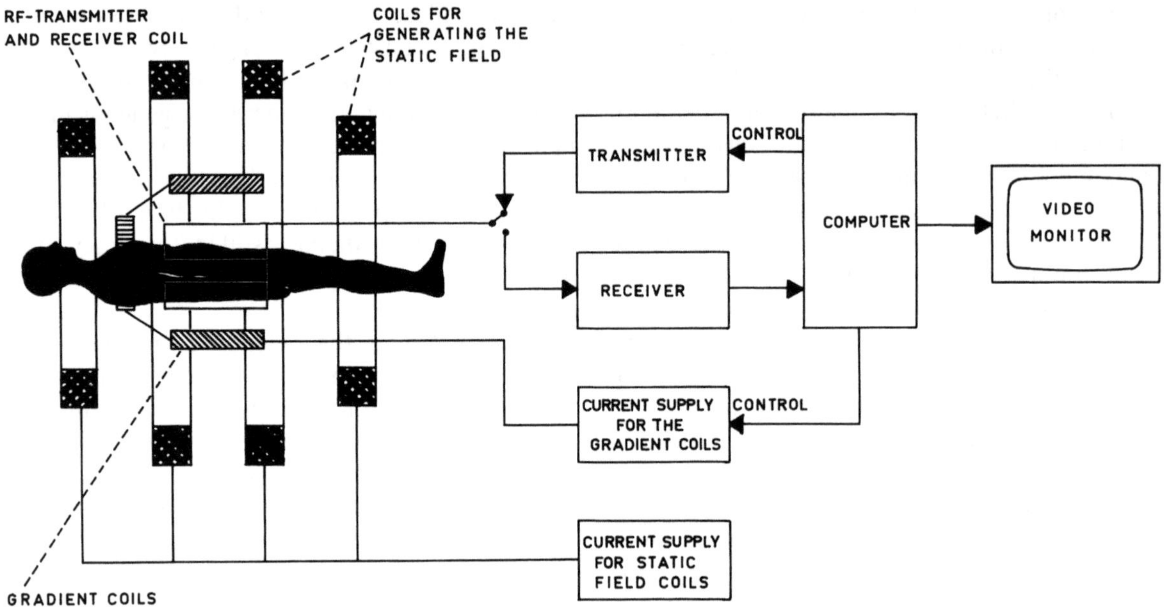

Fig. 1.5. Block diagram of a typical NMR imager. (After Habermehl and Graul 1982)

With the aid of specially arranged and tuned auxiliary coils (gradient coils) so-called gradient fields are produced which, individually, effect a linear modification of the homogeneous field in three orthogonal dimensions. With their aid, a spatial encoding of the received signal is made possible by switching them on in a given rhythm during the measurement. Their field strengths are about 1% of the strength of the static field; switching times of only a few milliseconds are required.

Via a saddle-shaped antenna that encloses the measuring volume the excitatory radio frequency field is emitted and the resonant signal of the precessing atomic nuclei is picked up. Accordingly, a transmitter and a receiver are alternately coupled with the antenna (Fig. 1.5). The closer the antenna to the patient, the larger is the signal received. For this reason, coils of different size are employed for whole-body and head examinations. Receiver coils in contact with the skin are able to "image" the underlying parts of the body with greater contrast.

In order to coordinate these basic elements of an NMR imager, a powerful small computer is required. This serves to control the gradient fields, the transmitter and the receiving apparatus via a quartz-controlled time base during measurement; it also stores the signals received in digital form prior to processing them in image construction.

As already mentioned, a nuclear induction experiment is made up of an excitation and a relaxation phase. In NMR tomography, too, all the picture points (pixels) must be measured individually. The result is a sequence of excitations and relaxations, in which only the size and polarity of the gradient fields change. In modern NMR imaging each excitation is followed by the measurement of a complete picture line, for which the mélange of frequencies brought about by the gradient field acting in the direction of the line is subjected to a Fourier transform analysis. With the spectrum obtained in this manner the spatial distribution of the signal strengths along the line is obtained directly. The time interval between two excitations is known as the repetition time, T. The product obtained by multiplying the line number Z and the repetition time T represents the minimum duration t required to produce a tomogram. Multiple measurements along a line with the aim of increasing the signal-to-noise ratio result in a multiplication of this measuring time with the number of signal averagings, n:

$$t = Z \times T \times n$$

If parallel sections are excited and measured in the time during which the previous slice is still in the recovery phase, during this time t, which is normally required to obtain a single tomogram, several slices can be obtained.

The angle of deflection of the magnetic moment relative to the direction of the static magnetic field (z direction) can be varied at will by the amplitude or the duration of the radio frequency excitation pulse. The greatest signal amplitude is obtained when the spins precess in the plane of the antenna – the xy plane. For the measurement, then, a so-called 90° excitation will, as far as possible, be employed. On the other hand by the use of a sequence of different excitations of the nuclear spin before the resonant signal is measured significantly differing image impressions can be obtained. This is an effect produced by the difference in emphasis of the various parameters – proton density, transverse and longitudinal relaxation times (Young et al. 1982; Edelstein et al. 1983).

The simplest measuring method is the repeated free induction-decay technique, during which for each line a 90° excitation is followed by a period of measurement (Fig. 1.6). Here, the spins precess in accordance with the respective relaxation times T_1 and T_2. Since the excitations are not line-selective – the entire sample volume is in resonance – the tissues relax, depending upon their specific longitudinal relaxation time T_1, to varying extents between the individual excitations and emit signals of varying intensity. Thus, in addition to the dependence of the measured value on the density of the hydrogen ions there is also a differentiation determined by the time constant T_1.

A more marked emphasis of the image contrasts by the longitudinal relaxation time T_1 is found in the inversion-recovery technique. Here, the described 90° excitation is preceded by a 180° pulse (inversion) imposed earlier by interpulse time τ. Tissues having T_1 times that are smaller than τ can adequately relax during this interval and emit a normal signal after the second excitation (Fig. 1.7). Longer relaxation times keep the direction of spin in inversion so that the prevailing situation is different when the 90° pulse is applied. This is recognizable on the basis of the phase relation of the signals. A sufficiently long repetition time ($T > 1$ s) ensures identical basic conditions for each line excitation. If the interval τ is chosen to be of the order of magnitude of two closely neighbouring relaxation times which are characteristic for various tissues, a clear separation of contrast is achieved. The inversion-recovery technique is particularly suitable for the visualization of parenchymatous organs such as the brain, the liver, the kidneys and the spleen.

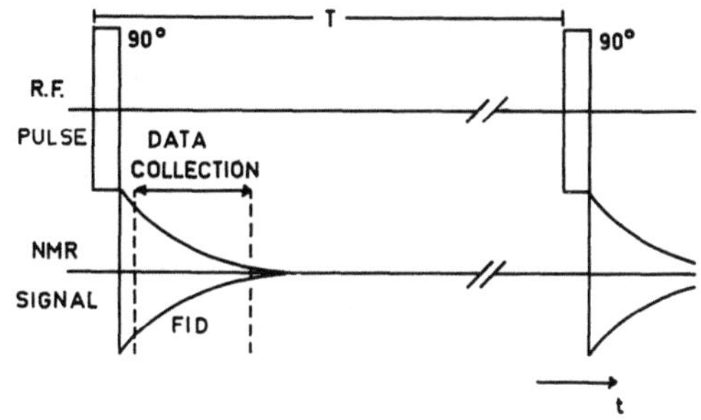

Fig. 1.6. Pulse and measuring sequence in the repeated free-induction decay technique

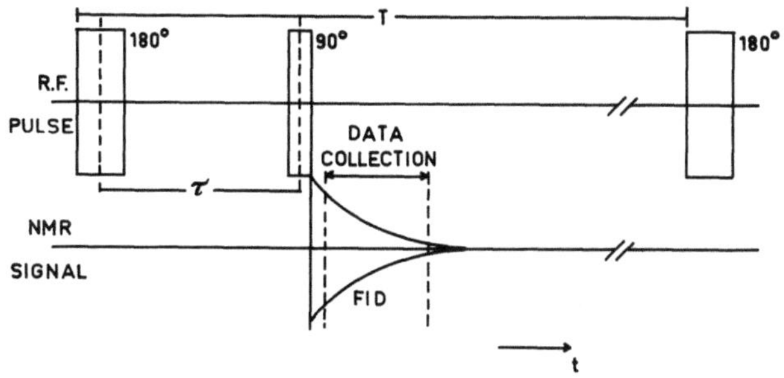

Fig. 1.7. Pulse and measuring sequence in the inversion-recovery technique

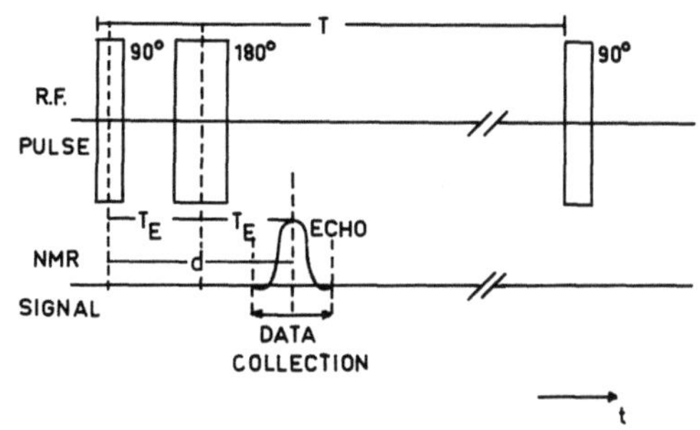

Fig. 1.8. Pulse and measuring sequence in the spin-echo technique

A third imaging technique is the spin-echo technique described as early as 1950 by Hahn; this results in an enhancement of the transverse relaxation time T_2. A 90° pulse is followed – after a time interval T_E, which is of the order of magnitude of the T_2 times – by a 180° excitation (Fig. 1.8). The spins precessing in the xy plane now undergo a reflection at the z axis, so that the spin directions moving out of phase by the T_2 relaxation now show a reversed behaviour and, after another interval T_E, again perform a synchronous precessional rotation (a large signal can be measured – called "echo"). The time elapsing between the excitation and the measurement of the signal ($2 \times T_E$) is called the read-out or echo delay d; together with the repetition time T it is the second important parameter in the spin-echo technique. A detailed discussion of this is presented in the next chapter.

An advantage of this technique, apart from T_2 enhancement, is an invariance of the picture signals vis-à-vis small-field inhomogeneities along with an exactly defined time window for the measurement. This so-called spin-echo imaging is used in place of the free-induction decay signal.

Owing to the large number of imaging techniques and instrument parameters and the representation of three different tissue characteristics, the procedure and operation of NMR tomography imaging appears highly complex and difficult to understand. For this reason, in the present initial clinical phase, efforts are being made to standardize the techniques and choice of parameters and to tailor them to the particular clinical problems. From this, useful practical work can quickly be expected while, at the same time, the possibility is preserved, by the use of special instrument settings, to arrive at a diagnosis of "problem cases" that are less amenable to the examination modalities that have been in use up till now (Crooks et al. 1982; Bydder et al. 1982; Bydder and Steiner 1982; Huk et al. 1983).

2. Contrast Behaviour

While the X-ray-based technique of CT visualizes only a single parameter, namely the density of the tissue, in NMR tomography imaging three properties of the tissues determine the image – the concentration of the hydrogen nuclei and the longitudinal and transverse relaxation times, T_1 and T_2. In the previous chapter it was shown that various imaging techniques permit an enhancement of the individual parameters. The information contained in the image is greatly expanded by this multiplicity, with the result that the assessment of the tomograms must be based on considerations other than those applicable to CT (Brant-Zawadzki et al. 1983). The present chapter serves as an introduction to the characteristics of the various imaging techniques and to the contrast attained with their use.

Spin-Echo Images

Spin-echo images obtained with short repetition times, exemplified by the images shown in the atlas section, are those most similar to X-ray CT scans with respect to their anatomical representation. A comparison of these two techniques, in particular with respect to the pixel numbers of characteristic tissues, will be of special interest to the reader with experience of CT.

Figure 2.1 shows a synopsis of the pixel numbers of different types of tissue in the head, the spin-echo values having been taken from the images presented in this atlas and the CT values of a modified visualization from the CT atlas by Gambarelli et al. (1977). In both cases the values are expressed in Hounsfield units; in NMR imaging, however, this unit does not represent attenuation values, nor does the zero correspond to the value for water, it has been used merely to facilitate a comparison of the two techniques. To date, no equivalent standardization has been established for NMR tomography imaging, since the variety of procedures and the possibility of selecting instrument parameters within wide ranges make it difficult to achieve a constancy of the measured values.

There are a number of clear differences between the two scales, which will immediately be apparent to the experienced reader. While in the case of CT three largely separate value (density) ranges can be observed, namely, air at the lower end of the scale, the soft tissues in the range between -120 and $+80$, and the bones from about $+200$ and above, the NMR values are uniformly distributed over the entire range of the Hounsfield scale. The soft tissues, which in the case of CT are concentrated in a narrow range, are spread over a wide range in NMR imaging, namely -800 to $+800$. This is also the explanation for the wide-open window setting of approximately 2000 channels, at which the images in the atlas were recorded from the video display.

The most important tissue parameters for spin-echo images with a short repetition time T are the density of hydrogen and the relaxation time, T_1. Owing to its high concentration of protons at short relaxation times fatty tissue is found at the upper end of the scale; this may be contrasted with its hypodense values in the CT technique. As a result of an almost complete lack of hydrogen nuclei bone and air both provide the faintest (smallest) signal. In the case of the remaining soft tissues the intensity of the signal can be observed to decrease with the decreasing amount of fat present. Epicranial aponeurosis (galea aponeurotica) as seen in the NMR image comprises subcutaneous fatty tissue, unspecific connective tissue and blood vessels and is thus to be found over a wide range of values between the remaining soft parts and the high values of fat. Bone marrow and connective tissue, too, contain varying percentages of fat and, on visualization, appear about as

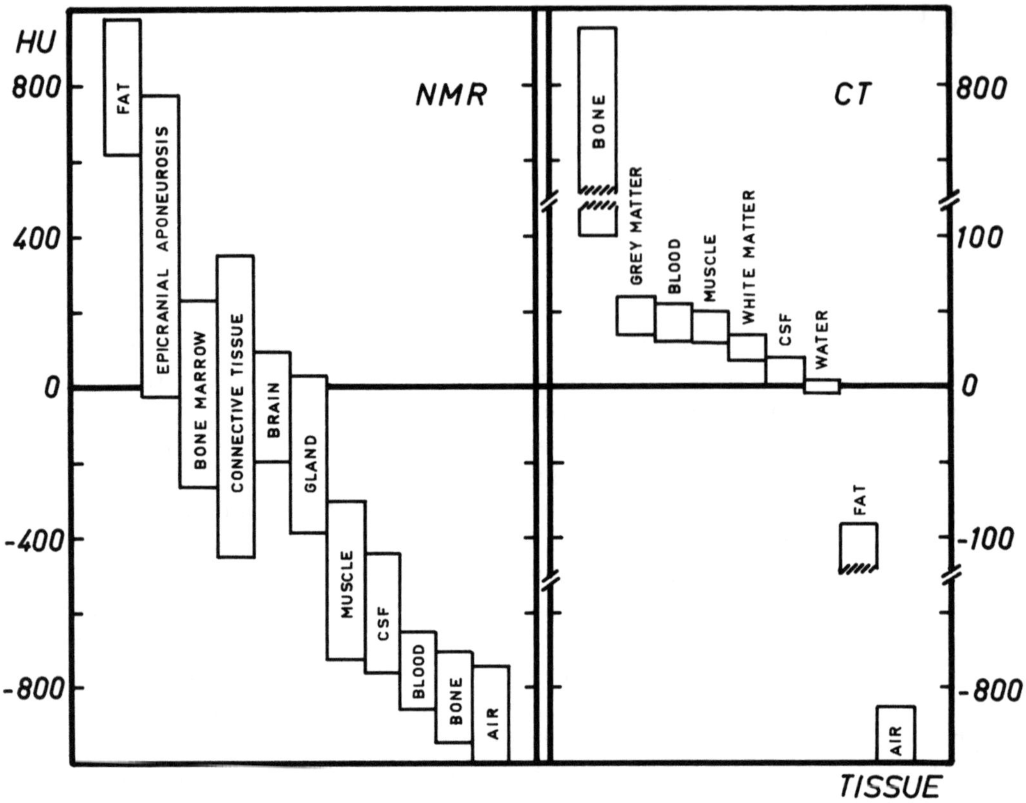

Fig. 2.1. Comparison of the pixel values for various tissues in spin-echo NMR ($T = 300$ ms, $d = 33$ ms) and CT scans, given in Hounsfield units

light as the tissue that is of particular interest – the brain. The latter has a proton content of about 10.5%, which is contained largely in the lipoproteins. With respect to these components, differences are observed between the grey and white matter of the brain. These would appear to be responsible for the differentiation described later. With little or no fat content, muscle tissue proves to be a relatively signal-poor substance. The values shown here are obtained mainly from the muscles of the tongue.

Quite a different behaviour can be observed on the part of the body fluids. Although these have a large proton content they are associated with very long relaxation times and therefore appear dark in this sequence. The slowest to relax is free water. Cerebrospinal fluid shows a somewhat faster relaxation behaviour, which is due to the "contamination" with salts and cellular elements; at appropriately short repetition times the intensity of signal it generates may be compared with that produced by muscle.

Substances in motion, that is, in the first instance the blood, must be considered separately. The contrast is additionally dependent upon the flow velocity orthogonal to the plane of the section (Crooks et al. 1982). The longitudinal relaxation time of blood is about 850 ms, so that blood vessels can be visualized, as light structures, only at longer repetition times. Blood vessels that are sectioned perpendicularly are always visualized only faintly with the spin-echo procedure, since the excited protons flow out of the measured slice in the interval between excitation and measurement. Differences in the visualization of muscle tissues, as can be seen in the images shown in the atlas, can be explained by the differences in their perfusion with blood, an effect that is of interest for measurement of blood flow through the capillary bed.

Variation of instrument parameters during imaging is, especially in the spin-echo technique, commonly employed. In particular the selection of various repetition times T (Chap. 1) for images of

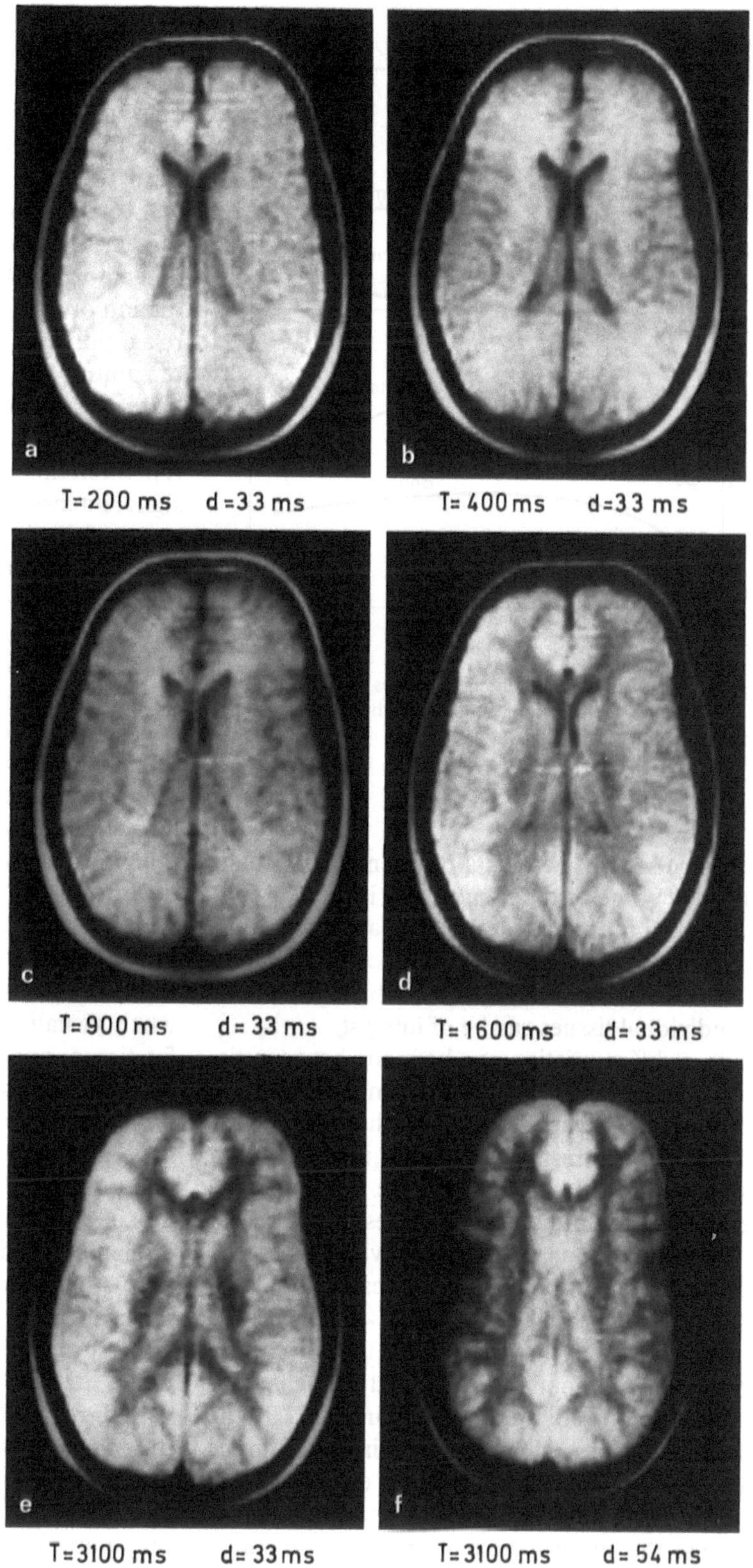

Fig. 2.2 a–f. Variation of the scanning parameters, repetition time, *T*, and delay, *d*, in the horizontal section H14

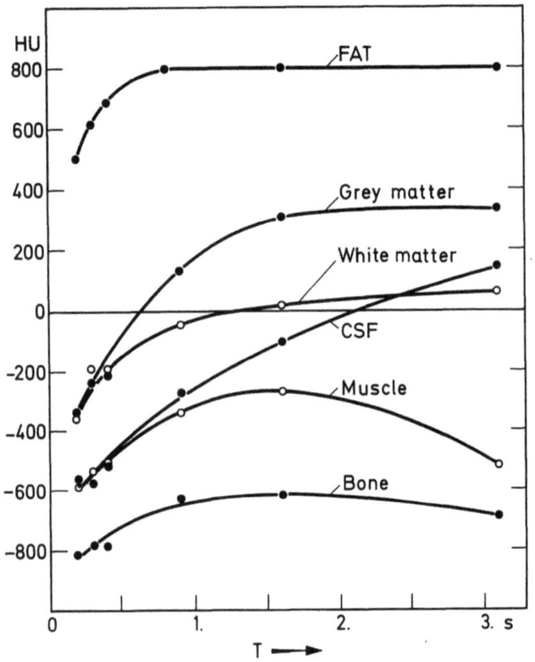

Fig. 2.3. The dependence of the pixel values of various tissues of the head on the repetition time, *T*, in the spin-echo technique (in Hounsfield units)

the same section permits a differentiated visualization of some of the tissues, and is thus of considerable importance for the evaluation of pathological structures. Even with respect to the purely anatomical visualization, the contrast behaviour of individual tissues can be of interest.

Thus, a differentiation can be made between tissues which remain largely unchanged when the parameters are varied and those that reveal clear changes in contrast. Of particular interest within the head are the substance of the brain and the cerebrospinal fluid of the ventricles, which reveal such an inconstant contrast behaviour when the parameter repetition time *T* and read-out delay *d* are varied. The image sequence in Fig. 2.2a–e clearly shows the changes in these tissues with increasing repetition time. The (contrast) behaviour is also graphically represented in Fig. 2.3, the measured values being obtained directly from the images. For the sake of clarity, errors are not shown. With rapidly repeated excitation ($T <$ 800 ms) the internal and external cerebrospinal fluid (CSF) spaces present a dark appearance, since the CSF is, as already mentioned, a fluid that is associated with long relaxation times.

As seen in the scale in Fig. 2.1, brain tissue is represented by an average image brightness and cannot be differentiated into white and grey matter. The subcutaneous fatty tissue surrounding the dark bony skull presents a particularly high contrast (light). From repetition times of about $T = 1600$ ms marked contrast shifts can be observed. The brain tissue now begins to differentiate into grey (light) and white matter (dark), a situation brought about by the differences in the T_2 relaxation times. The contrast relationships of the remaining soft tissues, such as muscles, glands, bone marrow, manifest little change, whereas the CSF produces a stronger signal with increasing repetition time, at values of $T = 3000$ ms, between signal intensities of the grey and white matter. The cause of this is the very long T_1 relaxation time in excess of 1500 ms, which makes possible a relaxation only with lengthy intervals between excitations.

The second parameter, the read-out delay *d*, also has an influence on the contrast. The delay between the 90° excitation pulse and the initiation of signal measurement corresponds closely with the transverse relaxation time, T_2. Tissues with a long transverse relaxation time appear lighter than those with a short T_2. Thus, by increasing the delay, long T_2 relaxation times can be enhanced and short relaxation times reduced with respect to image intensity. This effect can be seen in Fig. 2.2f: the scalp, which is usually very light when visualized, in particular the subcutaneous fat tissue, appears to produce almost no signal compared with the brain tissue, while the latter provides a somewhat clearer contrast between grey and white matter (corticomedullary contrast). A long delay is, however, associated with a loss of overall intensity, so that owing to the poorer signal-to-noise ratio a reduced spatial resolution is the result. This parameter is of interest in the evaluation of pathological tissue.

The particular advantage of the spin-echo technique are the imaging parameters which, when varied, result in a differing emphasis of the various tissues (Bailes et al. 1983). The sequence of images in Fig. 2.2a–f shows that marked differences occur only when the values of *T* and *d* are far apart. In clinical applications, therefore, at least four spin-echo images with differing imaging parameters are obtained of the sections of interest to the physician. For example, repetition times *T* of 300 and 1600 ms, and echo delays d of 33 and 66 ms are employed (Huk et al. 1983). Crooks et al. (1982) indicate a T of 500 and 1000 ms, and a d of 56 and 112 ms ($d/2$ of 28 and 56 ms). Experience shows that with such a variation, pathological tissues can manifest a greatly differing behaviour, which again contributes to their differentiation.

Images obtained with a short repetition time, T, and a short delay, d, provide excellent anatomical detail at short imaging times and are thus suitable as rapid orientating sections for a large measuring volume. Lesions with a very long (cysts, gliomas) or very short (lipoma) T_1 time and mass shifts are readily recognized. An increase in the delay at a rapid repetition time results in no further advantage, even in the evaluation of pathological findings (Huk et al. 1983). With longer repetition times, T, differentiation occurs in accordance with the T_2 relaxation time, and thus also the differences in contrast between white and grey brain matter are improved. Thus, lesions with demyelinating diseases can be detected (Young et al. 1983). An additionally increased delay provides good information in oedema of the brain, for example, in conjunction with an encephalitis. Disadvantages of the long repetition times are a long measuring time and poorer spatial and anatomical resolution, in particular of the internal CSF spaces.

T_1, T_2 Image

On the basis of the idealized assumption that for each pixel only one transverse and one longitudinal relaxation time applies, the signal values, S, of an NMR image can be computed with the aid of formulae which contain the instrument parameters T and d and the tissue-specific T_1 and T_2 times (Herfkens 1981). For spin-echo experiments, the following relation applies:

$$S \sim e^{-d/T_2}(1-e^{-T/T_1})$$

If three measurements are made of the same section, of which two have a varying parameter T at a fixed d, and two a varying d at fixed T, using a simple mathematical procedure the transverse and longitudinal relaxation times can be back-calculated for each pixel. From these data, another image can be obtained by assigning light grey tones to long relaxation times and dark grey tones to short relaxation times.

An example is shown in Fig. 2.4a–c, in which the slice corresponds to image H12 in the atlas sec-

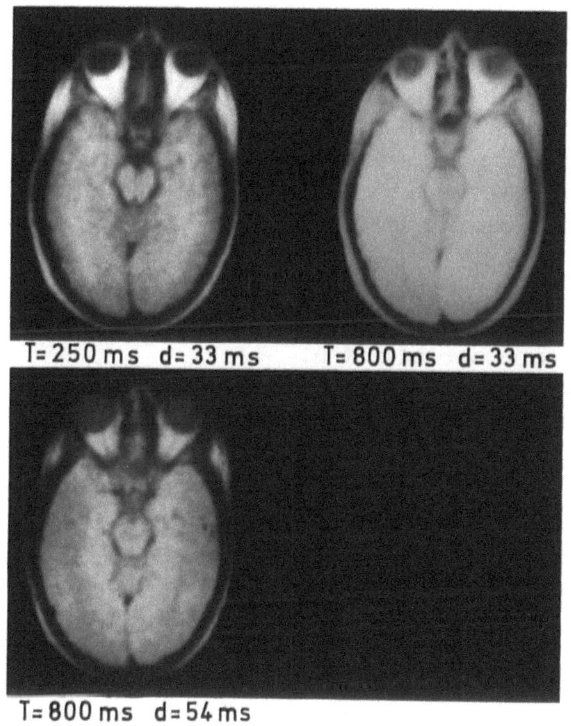

Fig. 2.4. (a) Images of the slice H12 at various repetition times, T, and delays, d, for the computation of T_1 and T_2 images

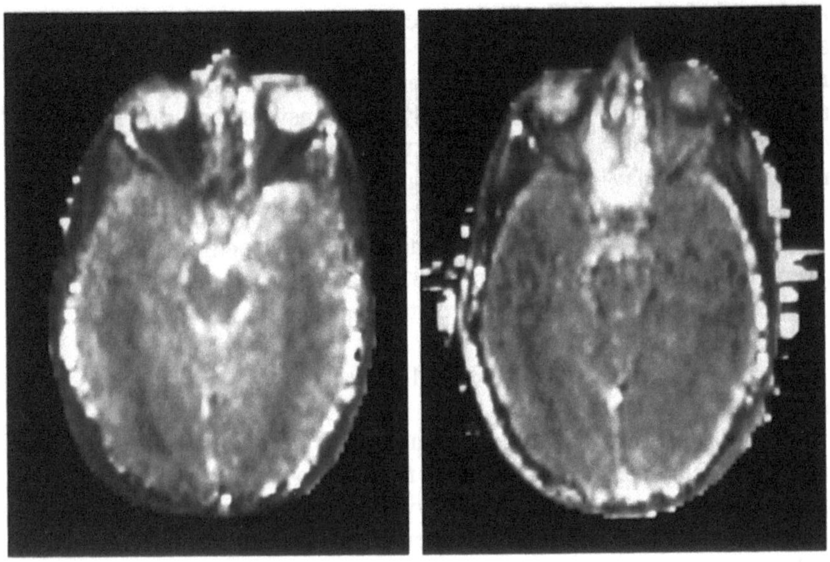

b c

Fig. 2.4. (b) T_1 image; (c) T_2 image

tion. Since for this series the amplification was kept constant the images clearly show the previously described differences in overall brightness, which increases with the lengthening of T and shortening of d. From these three images in Fig. 2.4a the representations of the spatial distribution of the T_1 and T_2 relaxation times shown in Fig. 2.4b, c were computed.

In the T_1 image the long relaxation times of the liquid in the anterior chamber of the eye and those of the basal cisternae surrounding the brain stem are represented in light shades of grey. The T_1 values lie between 1000 and 1800 ms. The retrobulbar fatty tissue presents a dark appearance, i.e a short T_1 – approximately 200 ms. The cerebral cortex and the medulla clearly show the T_1 differences, which in the inversion-recovery image provide the contrast.

In the T_2 image the tissues are visualized in a similar manner, although the differences are not so clear, in particular the transverse relaxation times of grey and white matter reveal much less contrast than would be expected from the spin-echo images with long repetition time. Table 2.1 shows, as examples, the repetition times of major tissues of the head. It must, however, be remembered that they change with the size of the static magnetic field applied. The values listed were measured at the field used here of 0.2 T (2 kG) (Huk et al. 1983).

This type of tissue visualization is certainly easier to interpret than in spin-echo images, but is strongly influenced by patient movements and equipment inconstancies. It can also represent a retrogression in tissue differentiation, since reciprocal parameter modifications do not find expression.

Table 2.1. T_1 and T_2 relaxation times of various tissues

Tissue	T_1 (ms)	T_2 (ms)
Fat	220/20	40/15
Muscle	400/20	50/15
White matter (medulla)	430/30	35/15
Grey matter (cortex)	640/30	45/15
Cerebrospinal fluid	1000–1800	30–50

Inversion-Recovery Image

In this technique the spins are excited twice at a time interval, τ. Tissues with relaxation times, T_1, greater than this interval produce a weaker echo than those with a shorter T_1. When these T_1 values are just over and just under the interpulse time τ the greatest differences in the signal are obtained. This is utilized in particular in the differentiation of the cortex and medulla of the brain. As can be seen in Table 2.1, the T_1 values for cortex and medulla are between 400 and 650 ms. A value of τ now in general use is 400 ms. Tissues with T_1 values of less than 577 ms ($= \tau/\ln 2$) appear light (positive on the Hounsfield scale), and tissues with longer values appear dark (negative). This results in a particularly good differentiation of the brain tissue and an unusually accurate impression of the anatomy of this organ. For the "constant" tissues, such as bone, muscle, glands, this imaging technique brings no advantages – quite the reverse, in fact, the spatial and contrast resolution are poorer, not least owing to the reduction in the signal-to-noise ratio. A scale of image values is superfluous for the inversion-recovery technique. For an evaluation it suffices to consider only four grey-scale ranges, to which the most important tissues are assignable:

Very light	fat
light	white brain matter (medulla)
grey	grey brain matter (cortex)
black	cerebrospinal fluid.

Figure 2.5 demonstrates the contrast pattern of an inversion-recovery image, taking the horizontal section H14 (p. 79) as an example. This may also be compared with the spin-echo variation – Fig. 2.3 – in which, at long repetition times, the cerebral cortex and the white matter are visualized in complementary contrasts.
The extremely clear visualization of the normal anatomy of the brain can also be utilized for the demonstration of pathological changes. Thus, for example, tumours – which are usually associated with a prolongation of the relaxation times – produce dark areas, while the surrounding oedema reveals only a slightly reduced signal intensity compared with normal tissue. Most of the substantial changes in the brain tissue are well represented with the inversion-recovery technique, while their differentiation is more readily possible using the spin-echo technique. In clinical applications the inversion-recovery image has become indispensable.

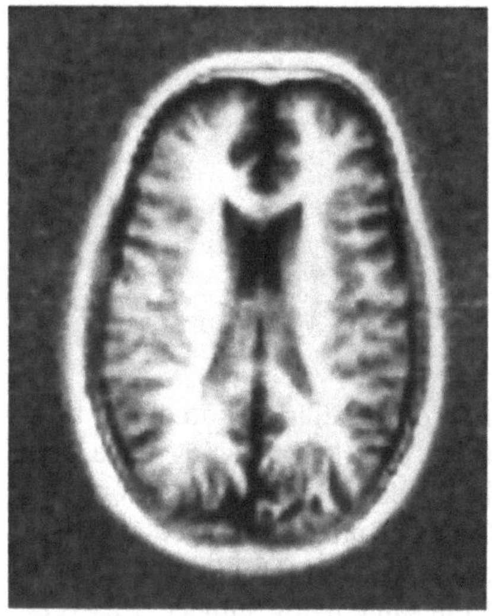

Fig. 2.5. Example of an inversion-recovery image with the instrument parameters repetition time $T = 1600$ ms and interpulse time $\tau = 400$ ms (equivalent to section H14a)

Thus, NMR imaging makes it possible, after an orientating examination of a body region, to make use of a rapid spin-echo sequence to provide an analysis of the pathological lesions topographically visualized by generating further images of the area of interest, employing varying imaging parameters.

3. Normal Anatomy of the Head As Seen in the NMR Image

In the production of the atlas section of this book all the advantages of NMR imaging have been utilized and visualized. The result is sectional images obtained in three planes arranged exactly at right angles to one another (Holland et al. 1980) of the head of the same living subject, a 23-year-old woman. The use of a living subject was justified both by the nonhazardous nature of the system (Budinger 1981) and by the fact that dead tissue sometimes displays a completely different imaging behaviour, so that a direct comparison with clinical images – the object of this undertaking – would not have been possible.

Although in the case of radiological CT the orbitomeatal plane has become accepted as the horizontal sectioning plane – which spares the relatively radiation-sensitive lens of the eye – here, orientation was to the "natural orthogonal coordinate system of the human head", which comprises the median sagittal plane, the plane of the face and the plane of the optic nerves. [The CA-CP (commissura anterior – commissura posterior) line, which is used in stereotactic surgery, lies approximately parallel to this plane.] These imaging planes are represented in Fig. 3.1 in relation to the median sagittal plane of the test subject. The dotted line represents the orbitomeatal plane, which forms an angle of about 15° with the plane of the optic nerve.

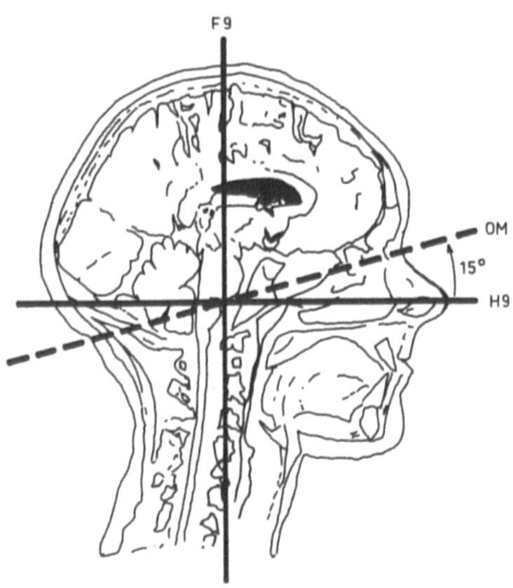

Fig. 3.1. Position of the imaging planes employed, represented in the median (sagittal) section S1 *(solid lines);* the orbitomeatal line *(OM)* is shown *dotted,* the point of intersection corresponds to the position of the external auditory meatus

Since all the images were obtained from the same head the imaging planes can be defined accurately vis-à-vis the relevant orthogonal planes. A grid that defines these planes is used to divide one-half of the head into $8 \times 16 \times 19 = 2432$ cubes, each side having a length of 1 cm, as shown in Fig. 3.2. This type of visualization makes it possible to assign each structure of the head to the coordinates of an orthogonal system, which is identified thus – S = sagittal, F = frontal, H = horizontal. The number adjacent to the letter identifies the respective section parallel to these planes. The numbering of the sagittal sections is from median to lateral (S1–S8), that of the frontal sections from ventral to dorsal (F1–F16) and that of the horizontal sections from caudal to cranial (H1–H19).

The pictorial section of this book is divided into three parts containing the images of the three orthogonal imaging planes – sagittal, frontal and horizontal. It contains a total of 43 images, each representing a different section. For each of the images showing brain structures the corresponding inversion-recovery scan, in which only the major anatomical landmarks are identified, is also provided.

The instrument parameters remained constant with a repetition time of $T = 300$ ms and an echo delay of $d = 33$ ms for all spin-echo images. A repetition time of 1600 ms and a interpulse time of 400 ms were used in the inversion-recovery technique. The window and level settings on the video display were selected to provide the best visual impression; this was obtained with a wide window of about 2000 channels and the level at middle range. For the inversion-recovery images the window was reduced to 1600 channels.

Each of the three sections is introduced by a brief description of the anatomical range shown. All the parallel sections of the relevant planes are represented by original pictures of orthogonal sections through the middle of the head.

Figure 3.3 shows the layout pattern of a page of the atlas with an inversion-recovery scan. In the lower right corner the designation of the imaging plane (S, F, H) together with the section number can be seen. To the left of this the former is shown in two orthogonal planes. The clinico-anatomical description of the slice position is indicated relative to the median sagittal section or external auditory meatus. Together with the spin-echo scan on this page, a good spatial impression is provided of the course of the section visualized. On the left-hand page the corresponding inversion-recovery image, which is provided with the index "a", can be seen. The text beneath this scan refers to the plane of the sections, i.e. it applies both to the inversion-recovery and the spin-echo scans.

On account of the wealth of contrast and clarity of the tomograms shown, the anatomical structures are labelled directly on the original image as it would be presented to the observer on the video screen of the evaluation unit. For the sake of clarity not all the structures visualized are identified; structures not labelled in one scan are identified in a neighbouring scan. All the anatomical structures are identified exclusively by means of the

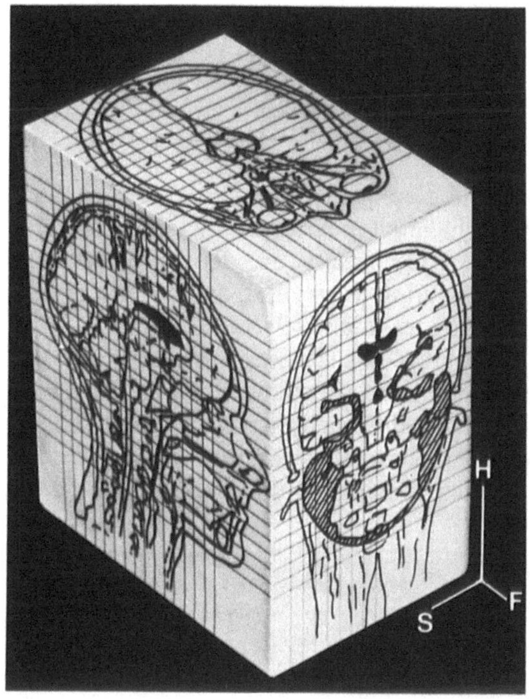

Fig. 3.2. The head of the subject divided up into 1 cm cubes by the sections of the three orthogonal planes S, F and H

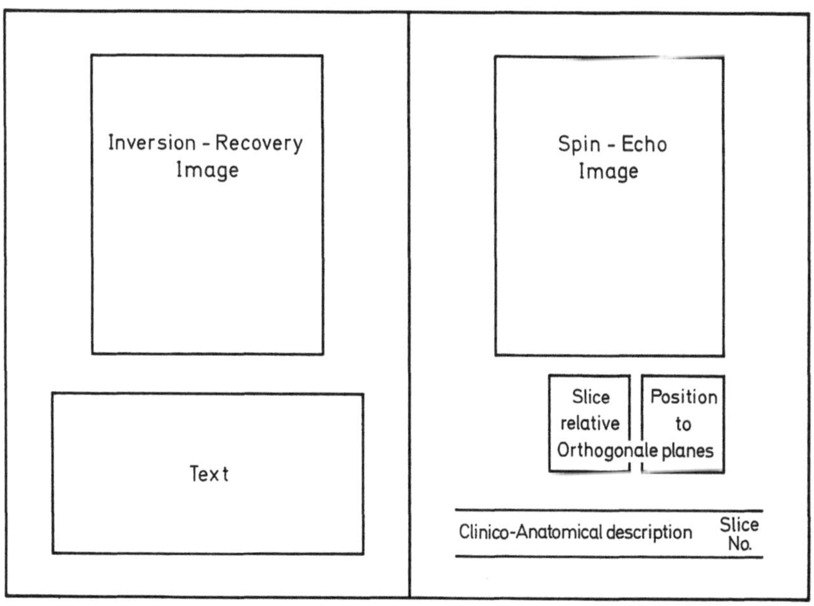

Fig. 3.3. Layout of a page of the atlas with an inversion-recovery scan

Latin nomenclature; where several terms are commonly employed for a given structure, corresponding references are found in the index. Here, in addition to the page number, reference is also made to the above-mentioned arrangement of the sections, so that the conception of the position and course of a given structure is facilitated for the viewer.

Fine Anatomical Structures

NMR tomographic imaging is excellently capable of reproducing topographical anatomical relationships. The contrasts are dependent both on the neighbouring tissues and, quite decisively, on the scanning technique selected.
The anatomical separation of the head into the neurocranium and the splanchnocranium is expedient for the following reasons. The brain tissue of the neurocranium reveals quite specific characteristics which require the utilization of the inversion-recovery technique and the spin-echo scans with a longer repetition time. As already mentioned the grey and white matter of the brain are differentiated through various mechanisms and thus permit unusually clear insights into the structure of the brain (see Chap. 2). In comparison with the substance of the brain, the CSF appears black in the inversion-recovery scan; in the spin-echo image, with a long T (approximately 2 s), the CSF produces a signal of about equal strength to that of the brain, which permits the visualization of the otherwise undifferentiable liquor-bone boundary.
With the exception of the eyeballs, to which the above remarks apply, in the region of the face tissues such as fat, muscles, connective tissues, glands and blood vessels predominate. Most of these structures have a relatively short relaxation time T_l and are thus visualized neither in the inversion-recovery scan nor in the spin-echo image with long repetition times to any great degree of contrast.

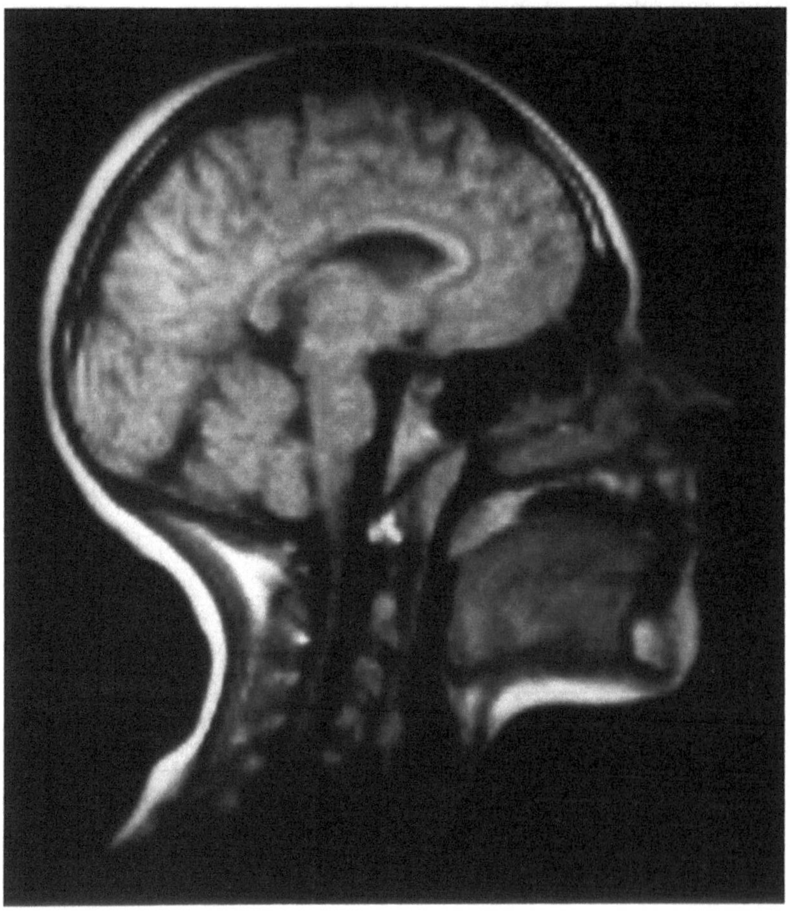

Fig. 3.4. The median sagittal section (S1) visualized with a repetition time of 300 ms and an echo delay of 33 ms in the spin-echo technique

The Neurocranium

The first orientating view is that of the median (sagittal) section, S1, which is represented in Fig. 3.4 at a repetition time of 300 ms. Clear topographical relationships can be observed from the spinal cord, via the midbrain, cerebellum and interbrain (diencephalon), to the cerebral cortex where, owing to the slice thickness of the scan (12 mm), the falx cerebri interferes with the convolutions and sulci of the medial surfaces of the hemispheres. Although the cerebral structures are readily distinguishable in all these scans they can best be followed in the inversion-recovery scan, S1 a. Here, in particular, the marginal portion of the gyrus cinguli may be mentioned, which runs in the cranial direction at a level of the posterior thalamus. It terminates the paracentral lobe dorsally and, at the same time, thus accurately defines the central sulcus. Depending upon the imaging plane, the convolutions and sulci are more or less clearly visualized. It goes without saying that the contrast to the external liquor spaces, blood vessels and venous sinus is greatest when the sectioning is perpendicular. Thus, for example, the precentral, postcentral and central sulci, which take a frontal direction, can be seen in the sagittal (S1–S7) and horizontal sections (H16–H19), but not in the frontal scans.

In the spin-echo images deeper layers of the brain present a relatively homogeneous appearance, which, however, is not at all the case in the inversion-recovery scans. Here, the clear visualization of the white matter results in an excellent delimitation of the convolutions, even when these are sectioned unfavourably.

The interbrain (diencephalon), which comprises the region from the crus cerebri to the gyrus cinguli, is particularly well visualized in the inversion-recovery technique. The basal ganglia are visualized at different intensities of contrast in the sections S1–S3, F6–F10 and H13–H15. The internal capsule, and in part also the external capsule, is visualized as an interposed light structure. Also in the spin-echo scan, H14, and the associated T variations, Fig. 2.2 a–f, the crus posterior capsulae internae is visualized as a contrasting dark area in the region of the interbrain.

While the nuclei of the hypothalamus region are not resolved, the hypophysis, together with the infundibulum and the surrounding sella turcica of the sphenoid can be distinguished particularly clearly at a repetition time of $T = 1600$ ms. In the detailed image of Fig. 3.5 even the anterior and posterior lobes of the pituitary (adenohypophysis and neurohypophysis) and the two bones that form the clivus, the sphenoid and occipital, are visualized.

The bony artefacts due to "beam hardening", which always interfere with the visualization of the posterior cranial fossa in CT imaging, do not occur in NMR images of this region (Young et al. 1981). All the structures of that region, the cerebellum, the medulla oblongata and the pons, are represented with uniform quality. They can be seen in the sections S1–S5, F9–F16 and H8–H11.

The midbrain, in particular the tectum and the pons, is well visualized in its gross anatomical structures, but the resolution of finer structures, such as the nucleus ruber, the substantia nigra and the nucleus olivaris, is inadequate. In the horizontal sections H9 and H12, although these structures are defined topographically, the contrast resolu-

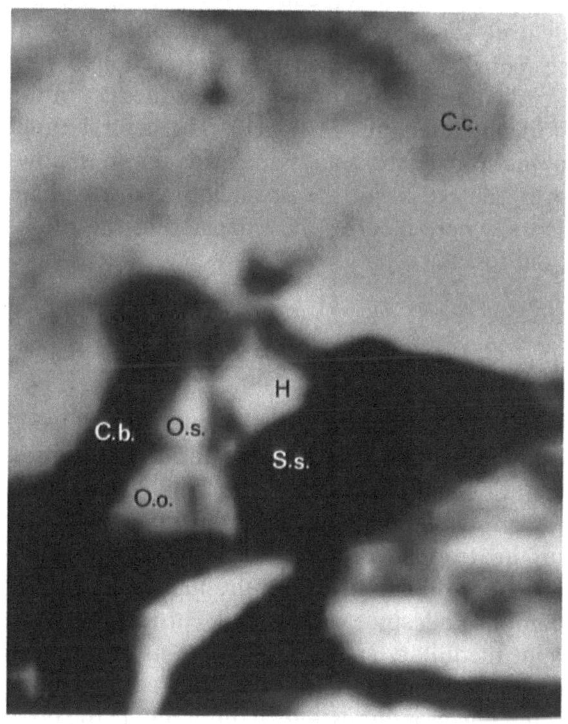

Fig. 3.5. Detailed visualization of the hypothalamus region and the pituitary gland in the median sagittal section S1 at a repetition time of $T = 1600$ ms. *C.b.:* Cisterna basilaris; *C.c.:* Corpus callosum; *H.:* Hypophysis; *O.o.:* Os occipitale; *O.s.:* Os sphenoidale; *S.s.:* Sinus sphenoidalis

tion is in this region not satisfactory. The substantia nigra should also be visible in H12.
The inversion-recovery technique clearly shows the difference in structure of the cerebellum vis-à-vis the cerebral cortex. In the horizontal sections H9–H11 the boundary of the occipital lobe and cerebellum can be distinguished much more readily than with the spin-echo image. The tentorium is sectioned obliquely and, as a result, loses its negative contrast through overlapping brain tissue. The dentate nucleus is represented in the sections S3a and H10a.
If we ignore the olfactory bulb only three of the cranial nerves can be clearly differentiated. While the optic nerve can be seen as a thick rope - like structure in a number of sections (S2, F2–5 and H12) and the trigeminal ganglion (in particular in the frontal sections) contrasts clearly with the CSF and bone, the oculomotor nerve (N III) is only just discernible in the cisterna basilaris, on a level with the crus cerebri, in section S2a.
The large CSF spaces are prominent in the atlas images. The ventricular system is more difficult to recognize in the region of the posterior and anterior horns and the aqueduct. In these regions the delicate ventricles are usually overlapped by brain structures, except when they are sectioned perpendicularly as, for example, the aqueduct in the horizontal sections H10–H12. The central canal of the spinal cord is not visualized here; it is, however, detected in the inversion-recovery scan F10a, where it forms the extension of the rhomboid fossa.

The Splanchnocranium

While the topography of the brain is relatively easy to distinguish in the scans, the differentiation of individual structures of the facial skeleton often presents something of a problem, in particular in the frontal sections. Even the anatomically most prominent structure, the mandible, proves complicated, since with its dark cortical bone and light medullary spaces it presents a rather inhomogeneous appearance. The thin cortical bone often makes it impossible to differentiate the bone marrow from neighbouring gland, fat or muscle tissue. In addition the ratio of cortical bone to bone marrow in the mandibular arch and the ascending rami varies, so that in particular in the case of the last mentioned the lateral pterygoid muscle can be confused with marrow.
The confusing multiplicity of muscles, fat, connective tissue and blood vessels can largely be unravelled with the aid of the above-mentioned spatial grid, which permits the position of each section to be determined relative to the other orthogonal slices, and thus previously defined structures can be detected and traced elsewhere. In this manner blood vessels in particular, such as the fine maxillary artery, the superficial temporal artery and the external jugular vein, can be defined. The large muscles of the neck and pharynx and the masticatory muscles can also be more easily identified using this aid. The most unequivocal topographic landmarks are first found in the horizontal sections and then the muscles in the frontal and sagittal sections are identified.
Other fine structures demonstrate the accuracy and fineness of the resolution in the NMR scan. In section F1 the lacrimal gland can be detected, while in the frontal sections F2 and F3 and in the horizontal section H12 the rectus muscles of the eye can be followed up to the annulus tendineus. In section H12a the anterior chamber of the eye and the lens are visualized. In sections H9 and H10 the nasolacrimal duct can be seen.
The horizontal sections H1–H5 in particular reveal astonishingly clear topographic relationships of the blood vessels and surroundings - without the use of contrast medium. The spinal cord can be recognized in the CSF-filled subarachnoidal space, surrounded by a tangle of ligaments, marrow and cortical bone of the vertebral arches.
Figure 3.6 represents the vertebral column and the larynx in a detailed presentation of the median sagittal section S1. Dorsally (on the left) the spinous processes of the cervical vertebrae, the white "rope" of the spinal cord within the dark spinal canal, the marrow-filled vertebral bodies surrounded by black cortical bone and, in between, the intervertebral discs, of which the pulp presents a light appearance, can be seen. Ventrally the light-coloured anterior straight muscles of the neck and the muscles of the pharynx are visualized. The air-filled hypopharynx and the larynx present a dark appearance, while the epiglottis can clearly be seen projecting into the pharynx.
The various sections of the larynx can be recognized with ease: from cranial to caudal - the introitus, a light transverse structure, the plicae ves-

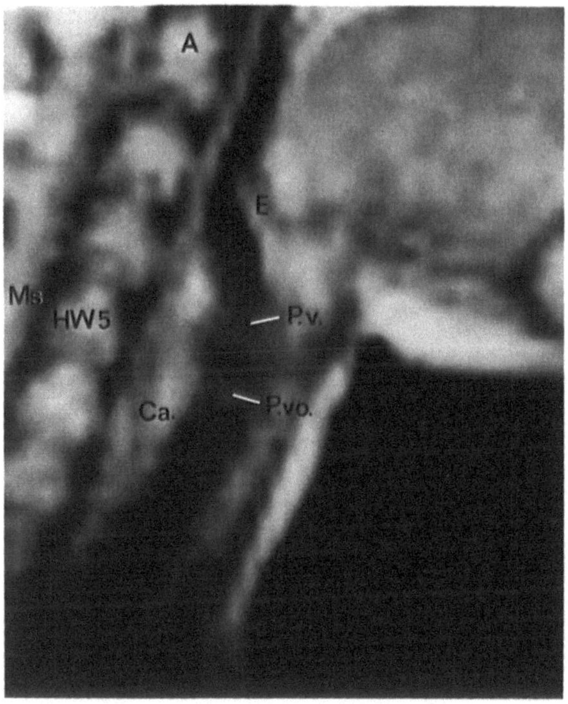

Fig. 3.6. Detailed visualization of the hypopharynx and larynx region in the median sagittal section S1 ($T = 300$ ms, $d = 33$ ms). *A.:* Axis; *Ca.:* Cartilago cricoidea; *E.:* Epiglottis; *HW5:* fifth cervical vertebrale; *M.s.:* Medulla spinalis; *P. v.:* Plicae vestibulares; *P. vo.:* Plicae vocales

tibulares and the adjacent ventriculus laryngis, delimited caudally by the barely visible plicae vocales, can be distinguished. The transition to the trachea, which is visualized ventral to the fifth cervical vertebra as a black (air-filled) structure, is on a level with the cricoid cartilage. In the frontal (F7–F9) and horizontal sections (introitus H1) a good, highly contrasted visualization of this area is found. Tumours and infiltrative growths in this region should be detected without any difficulty.

The list of the identifiable fine structures could be continued; in the atlas section more details are presented in the descriptions of the relevant sections. All in all, some 250 different structures in the head and neck areas are identified – proof enough of the excellent reproduction of anatomical details that this imaging system makes possible.

4. Sagittal Sections

Scanning was started with the median sagittal section, and this was followed by a sequence of scans obtained to the left of the median sagittal plane at intervals of 1 cm. The selected section thickness was 12 mm. The eight sagittal sections obtained in this manner cover the whole of one-half of the head extending from the median line to the left external ear. Corresponding inversion-recovery images comprise the sections S1–S4; slice thickness was 11 mm. Figures 4.1 and 4.2 show the direction and position of all eight slices (S1–S8) with reference to the middle sections of the two orthogonal planes (H12, F1/F9). The lines in the original scans represent the idealized middle plane of each section.

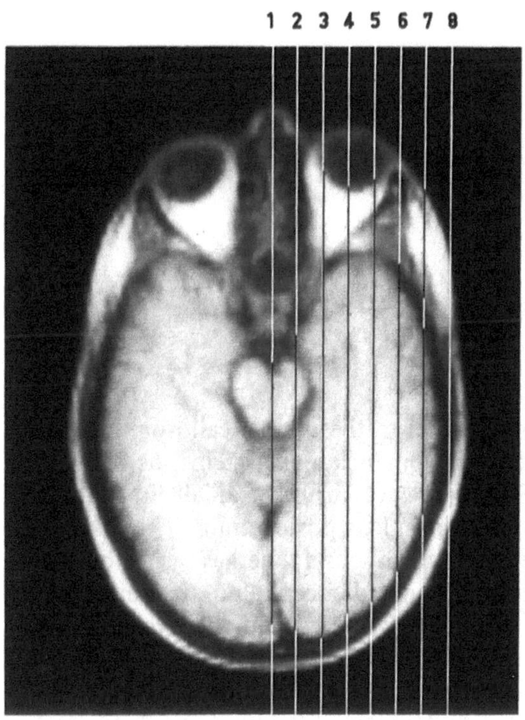

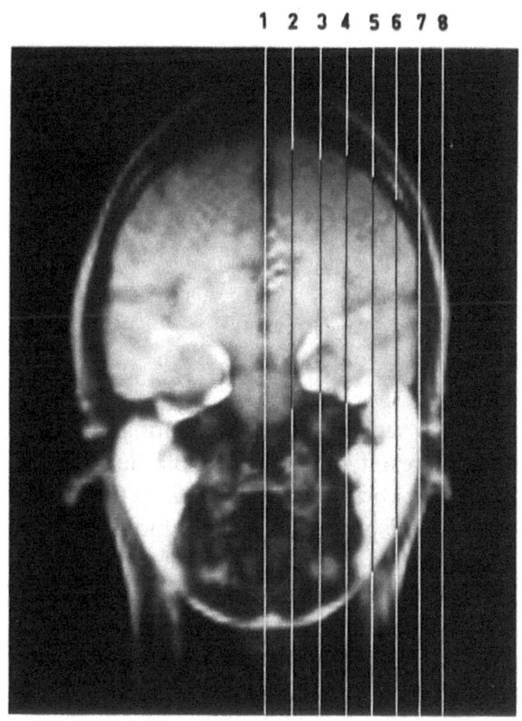

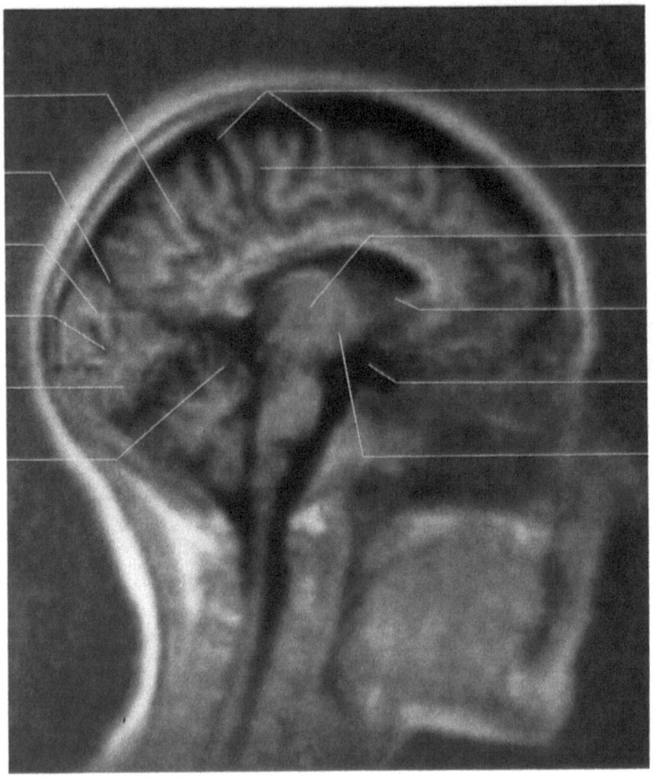

Fig. S1 a

S1 is a median sagittal section presenting the right aspect; as a result of the width of the slice of 12 mm the falx cerebri is concealed by the structures of the hemisphere. The convolutions and fissures (sulci) in the medial surface of the cerebral cortex can be recognized; they are particularly well discernible in the inversion-recovery scan. Here, the main features visualized are the sulcus parieto-occipitalis, the line of separation between the parietal and occipital lobes, the sulcus calcarinus, representing the topographic landmark of the visual field, and the gyrus cinguli, which follows a course parallel to the corpus callosum and, in a number of nomenclature systems, as shown in Fig. S1 a, curves off in the cranial direction at a level with the sulcus postcentralis. The third ventricle is too narrow and "merges" with the tissue of the interbrain (diencephalon, thalamencephalon), with the thalamus and the hypothalamus. The hypophysis can be seen, without the infundibulum, "nestling" in the light sphenoid bone (sella). A better view of the pituitary gland can be found in Fig. 3.5. A comparable section is illustrated in Fig. 3.4.

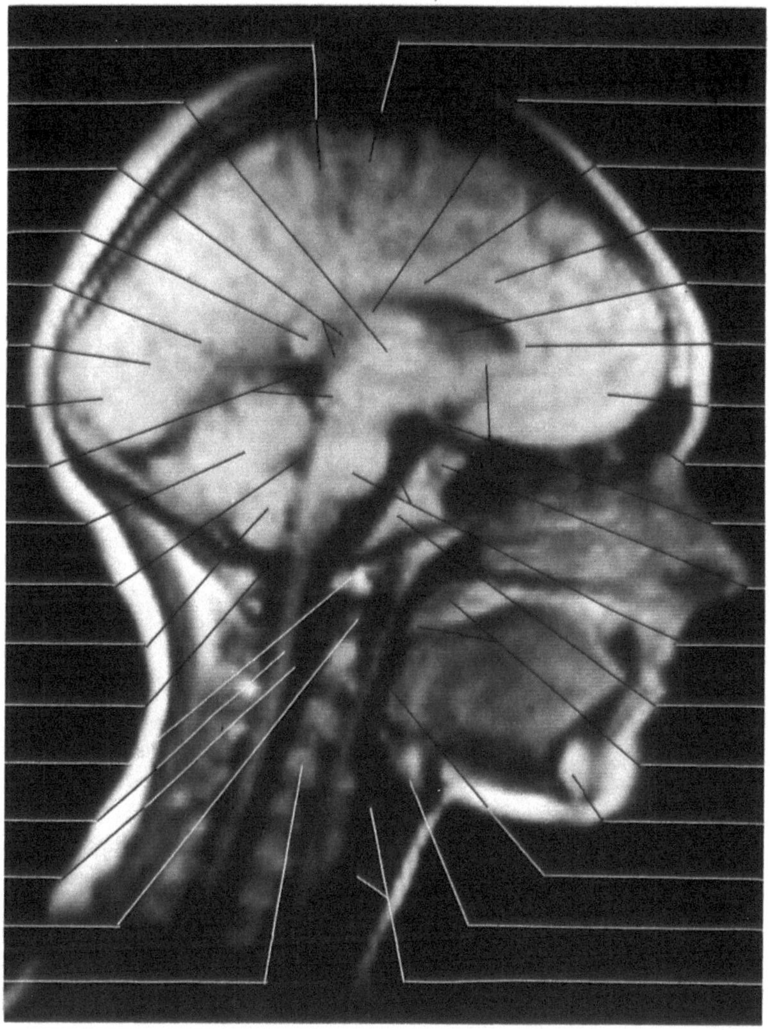
Sulcus centralis
Thalamus
Commissura posterior, Corpus pineale
Splenium corporis callosi
S. parietooccipitalis
Cuneus
S. calcarinus
Cist. ven. cereb. magn., Lamina tecti
Vermis cerebelli
Ventriculus IV
Tonsilla cerebelli
Cisterna magna
Atlas
Medulla spinalis
Cavum subarachnoidale
Dens axis
Corpus vertebrae
Gyrus praecentralis
Corpus fornicis
Corpus callosum
Gyrus cinguli
Ventriculus lateralis, Septum pellucidum
Genu corporis callosi
Gyrus fontalis sup.
Sinus frontalis
Commissura anterior, Chiasma nervi optici
Hypophysis
Pons, Cisterna pontis
Os sphenoidale, Clivus
Palatum molle, Uvula
Mandibula
Hypopharynx
Epiglottis, Corp. adiposum laryngis
Larynx, Trachea

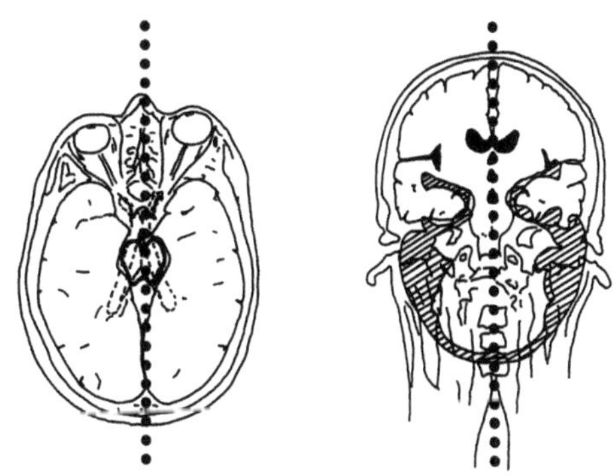

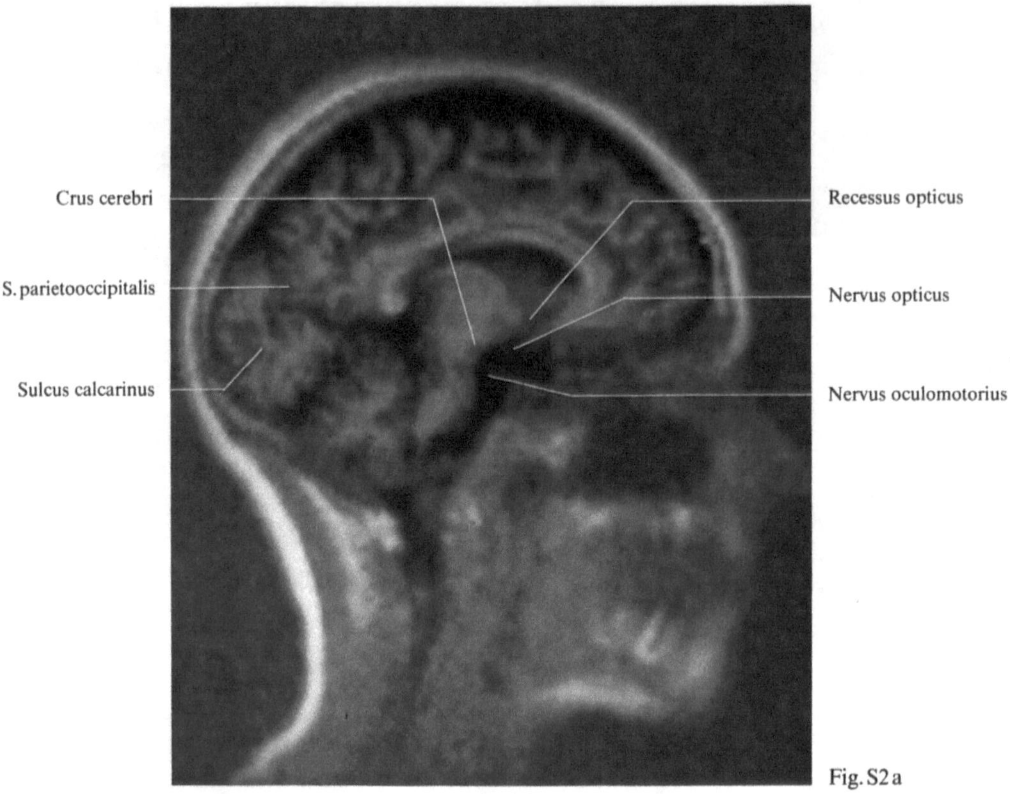

Fig. S2 a

Section S2 shows virtually the same structures as S1. The lateral position of the section can be most clearly recognized in the region of the pons and in the more caudal area. In the inversion-recovery image, Fig. S2 a, the third cranial nerve (nervus oculomotorius) is just discernible in the cisterna basilaris. The lobulus paracentralis, site of the primary motor and sensory cortical areas, is more clearly visualized, cranial to the thalamus, than in section S1. Through the tissue of the tongue the first signs of the lateral teeth can just be made out.

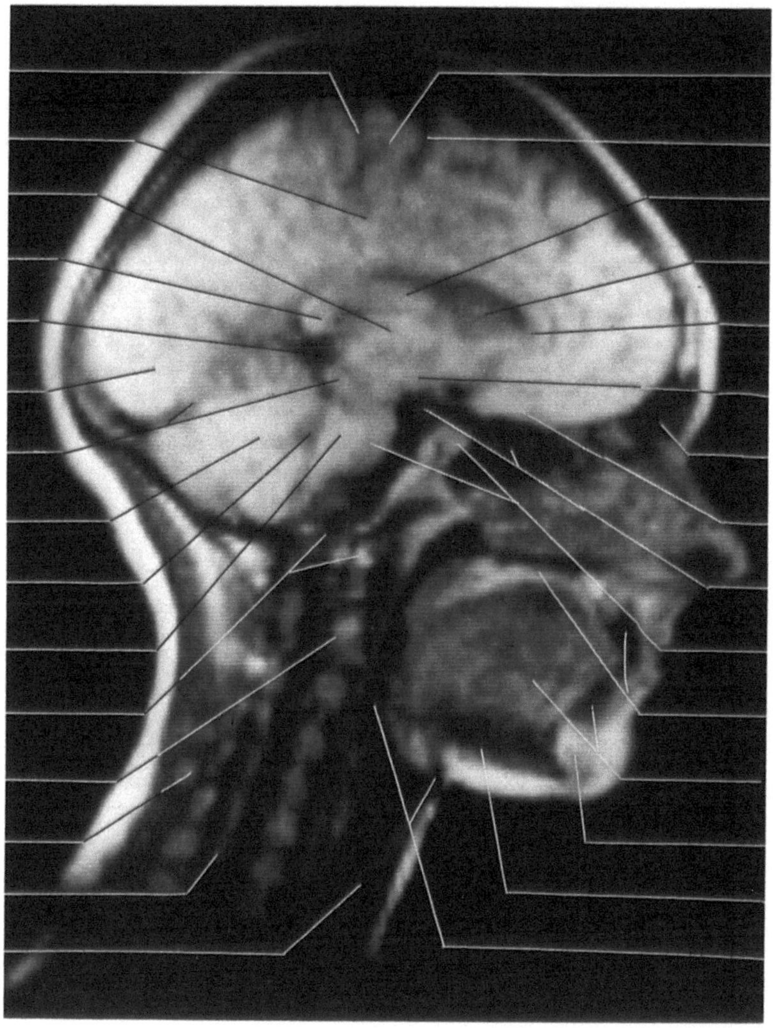
Sulcus centralis
Gyrus cinguli
Thalamus
Splenium corporis callosi
Cisterna venae cerebri magnae
Lobus occipitalis
Lamina tecti, Tentorium cerebelli
Cerebellum
Ventriculus IV
Pedunculus cerebelli medius
Foramen magnum, Atlas
Axis
M. semispinalis capitis
Medulla spinalis
Trachea
Gyrus praecentralis
S. praecentralis
Corpus fornicis
Ventriculus lateralis
Genu corporis callosi
Hypothalamus
Sinus frontalis
Gyrus rectus
Corpus mamillare, Sinus sphenoidalis
Hyophysis, Pons
Palatum durum, Dentes superiores
Lingua, Dentes inferiores
Mandibula
M. digastricus
Hypopharynx, Os hyoideum

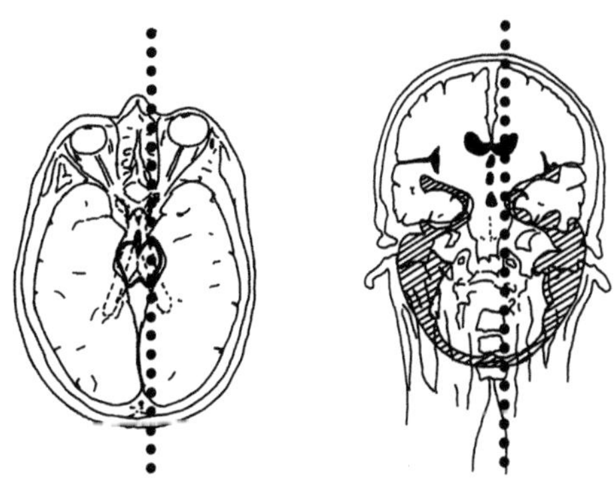

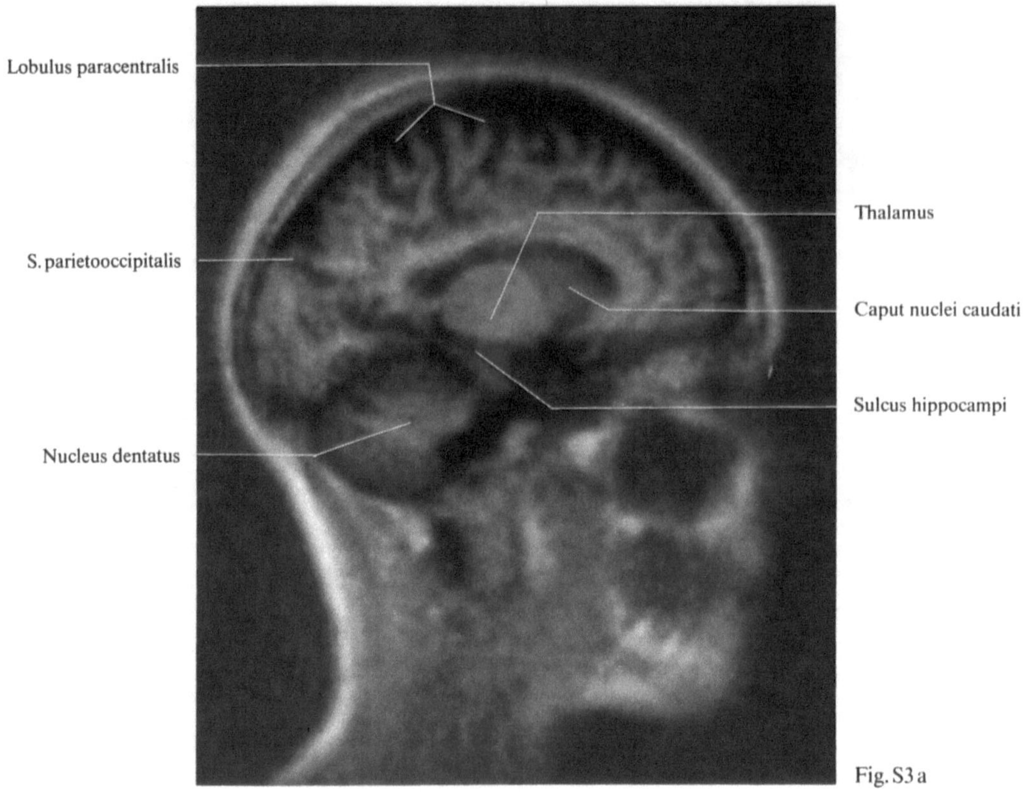

Fig. S3 a

Section S3 lies along the line joining the medial rectus muscle of the eye to the uncus gyri parahippocampalis. The curved lateral (left) ventricle arches anteriorly over the caput nuclei caudati and posteriorly over the thalamus. These structures are well demarcated in Fig. S3 a, the inversion-recovery scan. In the cerebellum formations such as the nucleus dentatus can be seen, while the spin-echo scan presents a virtually homogeneous aspect. In the facial region the medial portion of the masticatory musculature, the musculus pterygoideus medius and the teeth of the left side can be recognized.

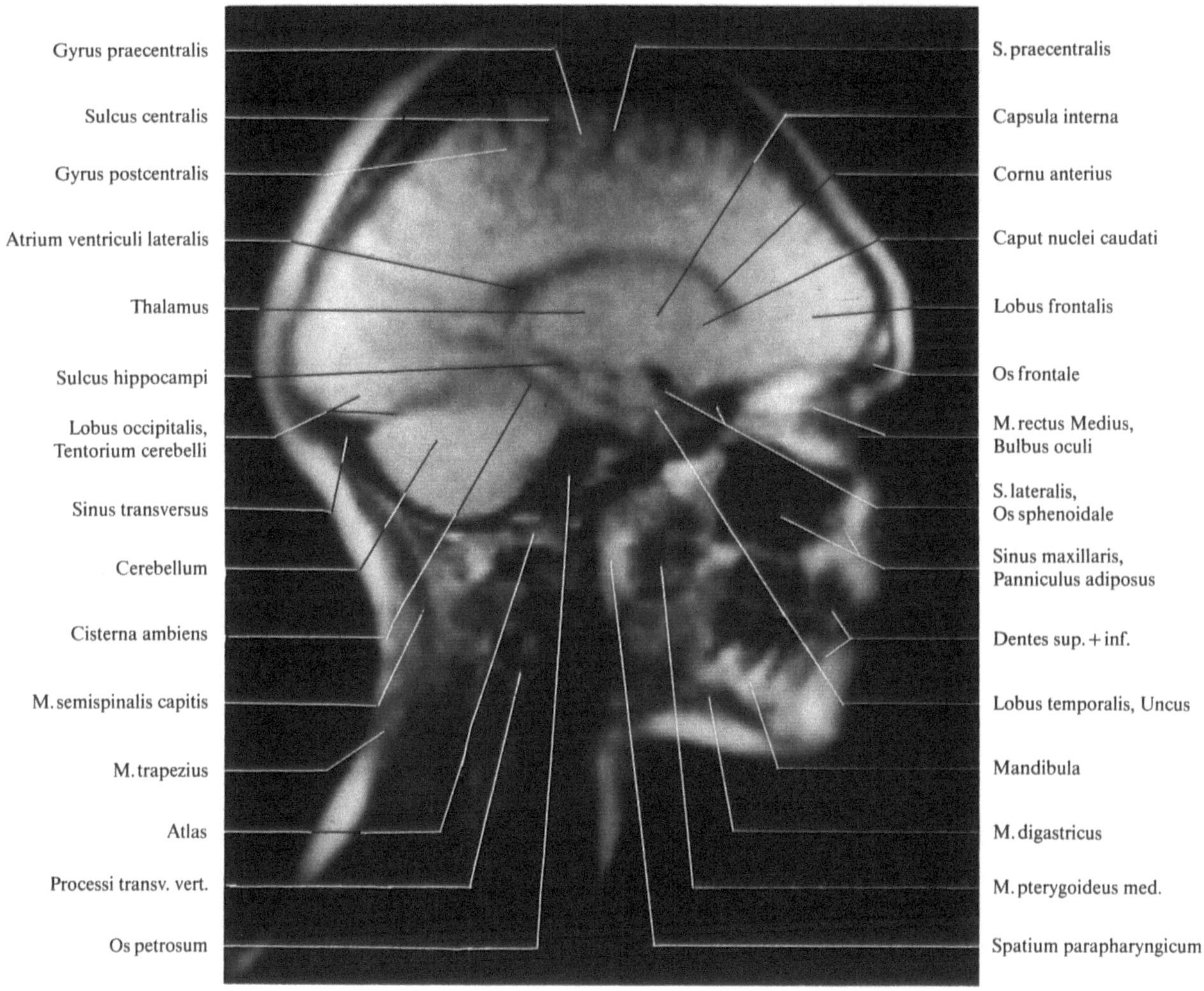

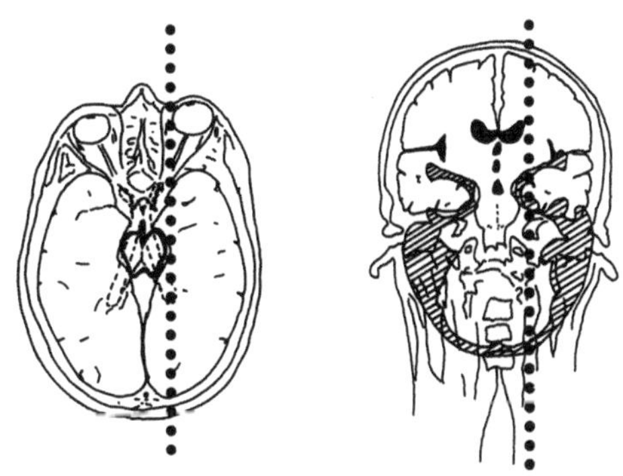

Sagittal section 2 cm lateral to the midline **S 3**

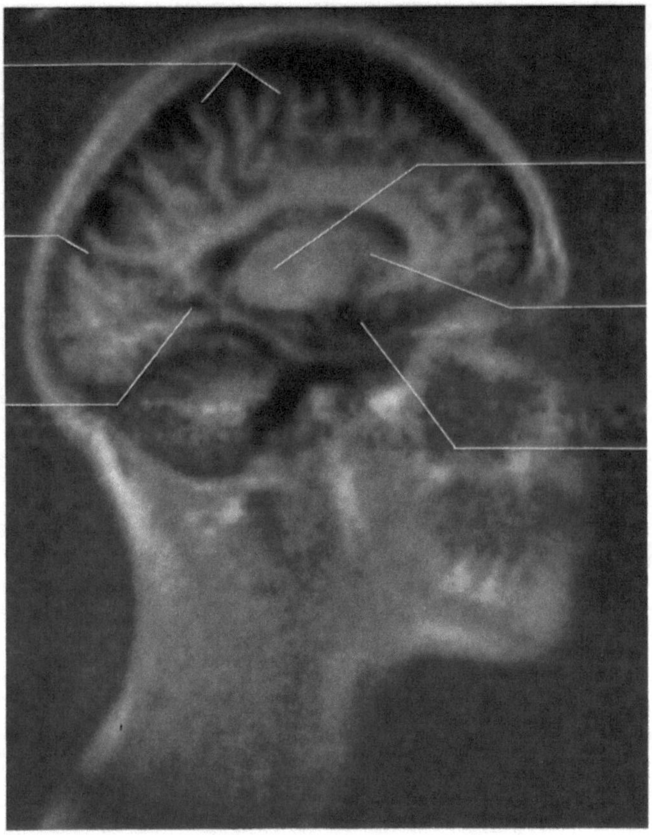

Fig. S4a

Section S4 passes through the medial parts of the eye and the cornu inferius (inferior horn) of the lateral ventricle. The temporal lobe is separated, optically, from the diencephalon by the sulcus lateralis. The inversion-recovery scan S4a is somewhat more medial, so that in place of the nucleus lentiformis the thalamus appears too large. In the ventral part of the neck the arteria carotis communis and the vena jugularis interna can be recognized as black, vertical ropelike structures.

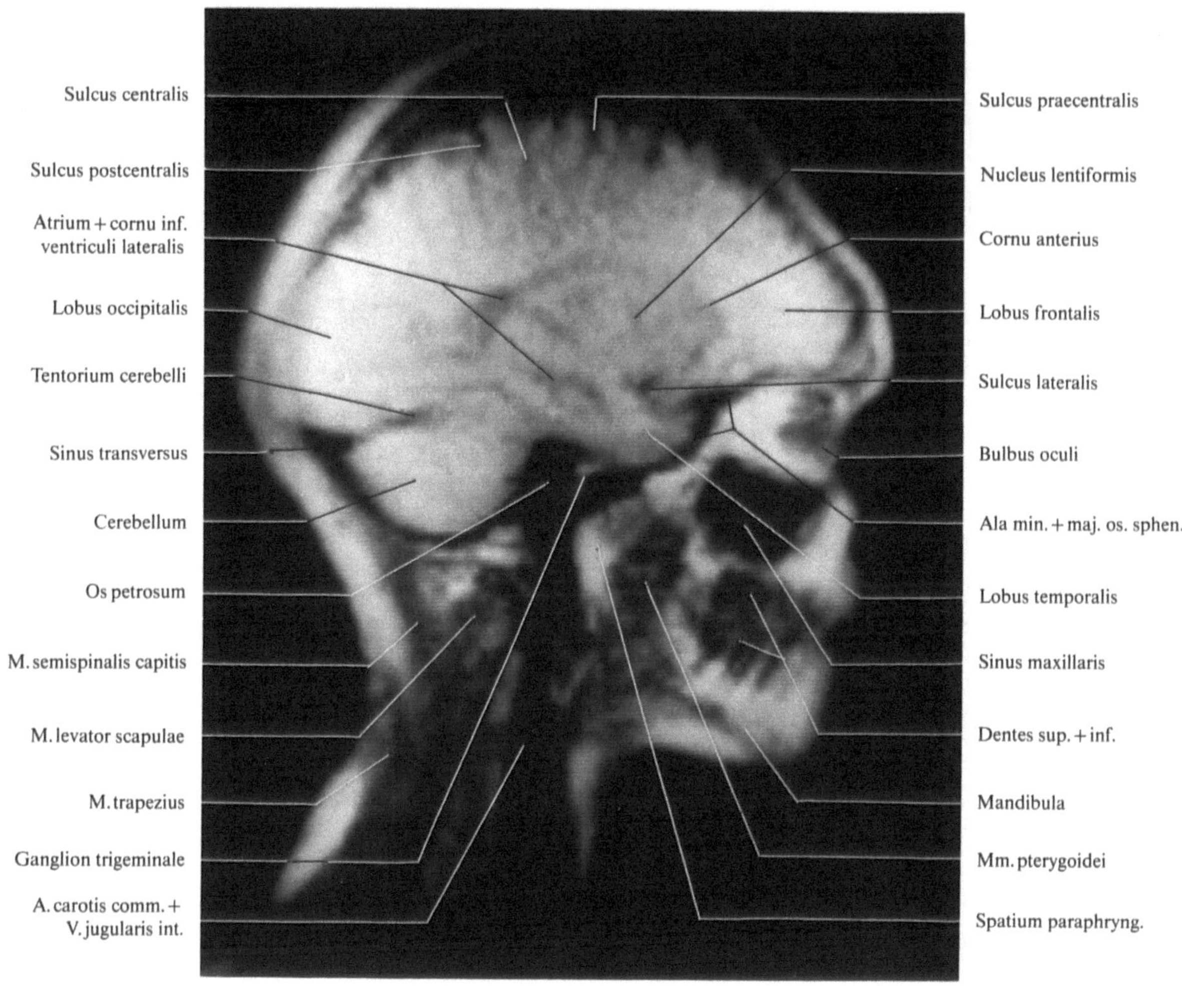
Sulcus centralis
Sulcus postcentralis
Atrium + cornu inf. ventriculi lateralis
Lobus occipitalis
Tentorium cerebelli
Sinus transversus
Cerebellum
Os petrosum
M. semispinalis capitis
M. levator scapulae
M. trapezius
Ganglion trigeminale
A. carotis comm. + V. jugularis int.
Sulcus praecentralis
Nucleus lentiformis
Cornu anterius
Lobus frontalis
Sulcus lateralis
Bulbus oculi
Ala min. + maj. os. sphen.
Lobus temporalis
Sinus maxillaris
Dentes sup. + inf.
Mandibula
Mm. pterygoidei
Spatium paraphryng.

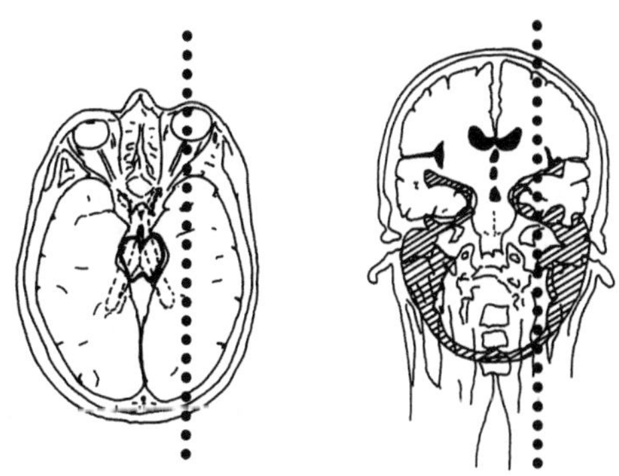

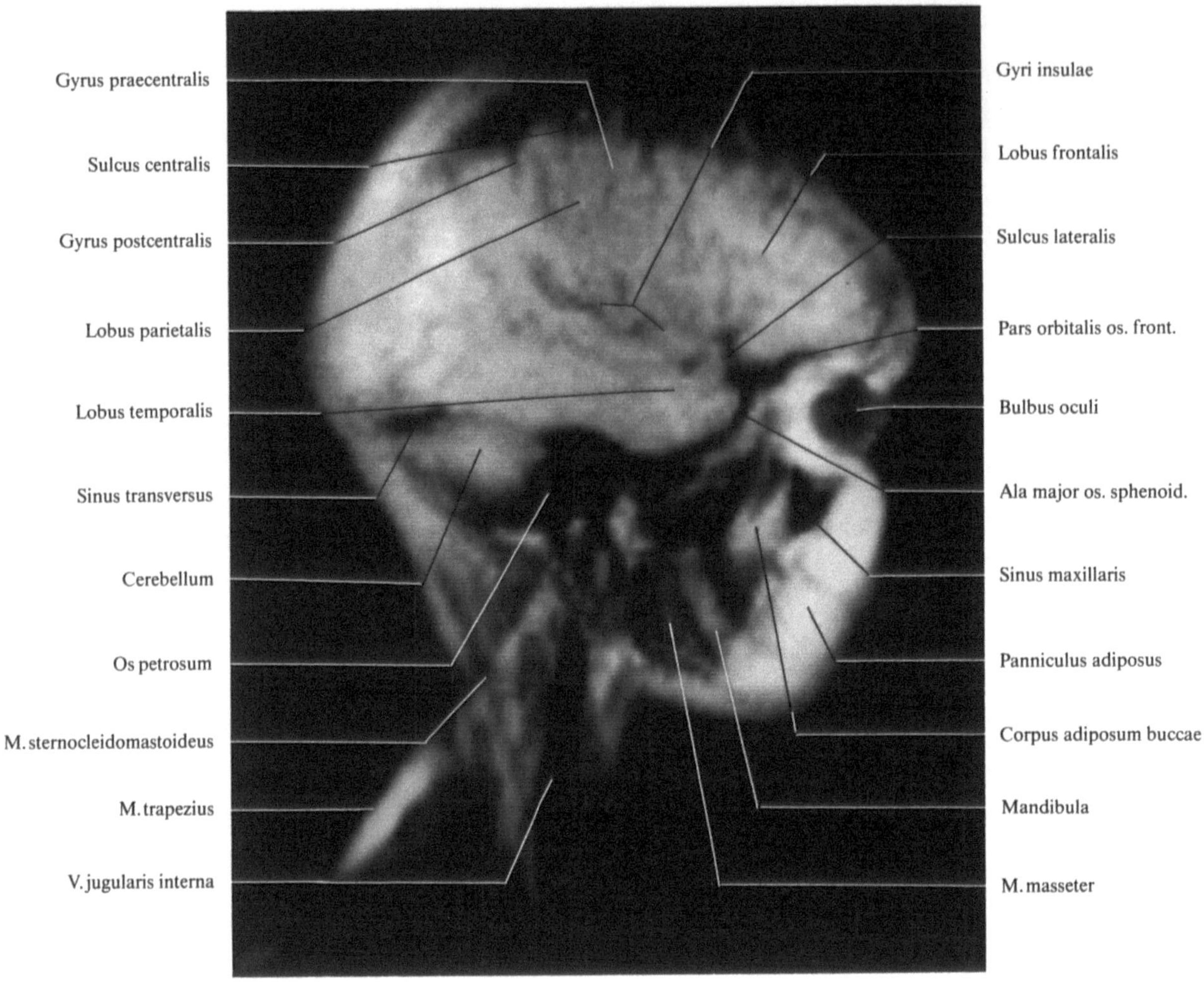

Section S5 is located lateral to the middle of the eye; the lens is no longer visualized. Well visualized structures of the brain are the lobulus paracentralis and the gyri insulae. The parietal lobe is shown in its full length. The sinus transversus is readily recognized as a dark area separating the occipital lobe from the cerebellum. In the facial region the very light buccal fat (panniculus adiposus) is better visualized. Caudal to the corpus adiposum buccae the ramus (ascending branch) of the mandible becomes evident as a bright medullary area framed by black cortical bone. In the region of the neck the internal jugular vein is represented medial to the musculus sternocleidomastoideus.

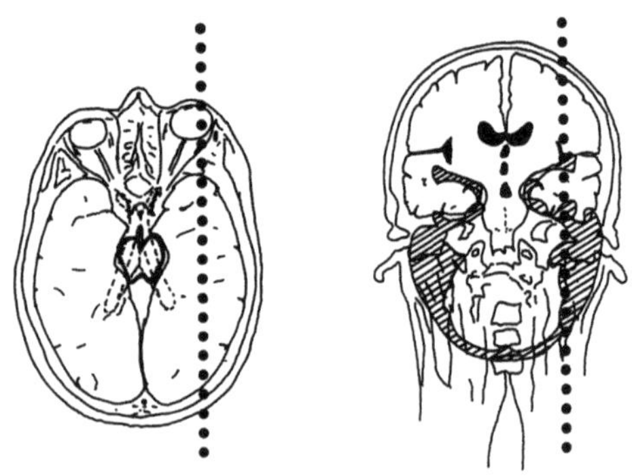

Sagittal section 4 cm lateral to the midline **S 5**

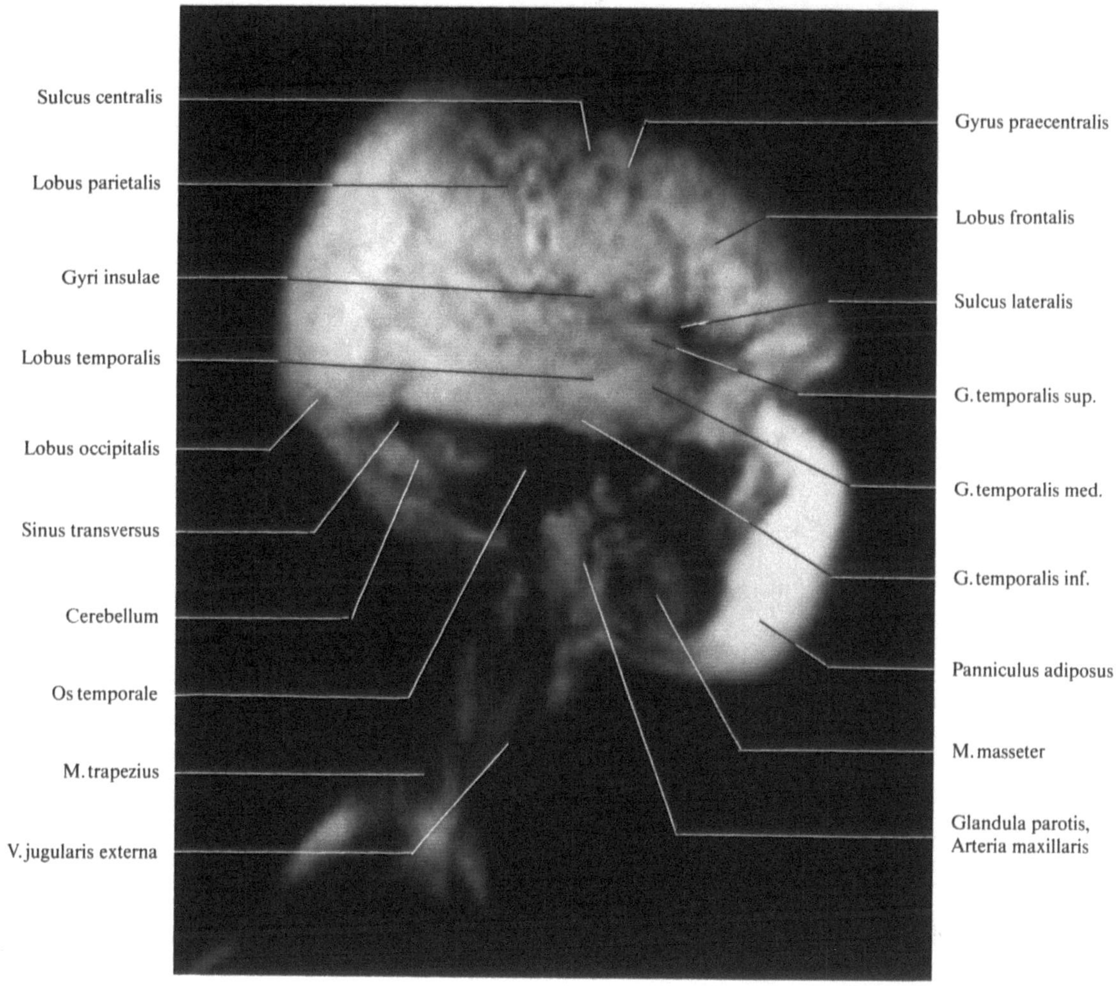

Section S6 passes through the lateral corner of the eye. The structures of the brain now become less readily distinguishable, only the sulcus lateralis is easily identifiable, while the lobulus paracentralis can only just be discerned. In the region of the jaw the parotid gland can be recognized dorsal to the musculus masseter; at its medial border the arteria maxillaris appears as a dark, elongated ribbon. Caudally the only structure of the neck to be visualized is the musculus sternocleidomastoideus, together with the tortuous vena jugularis externa. In the horizontal sections H1–H4 the course of this vein can readily be observed.

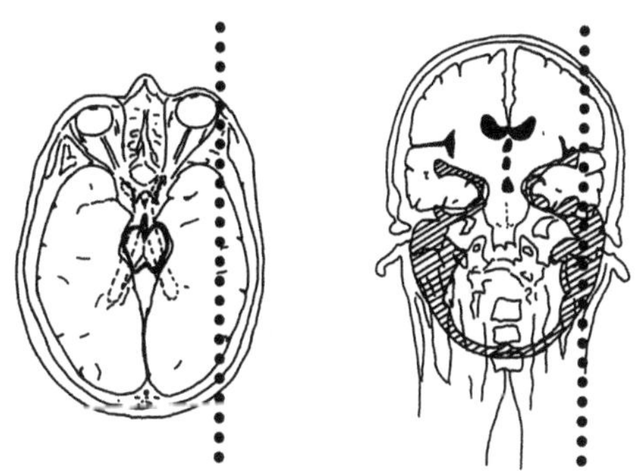

Sagittal section 5 cm lateral to the midline **S 6**

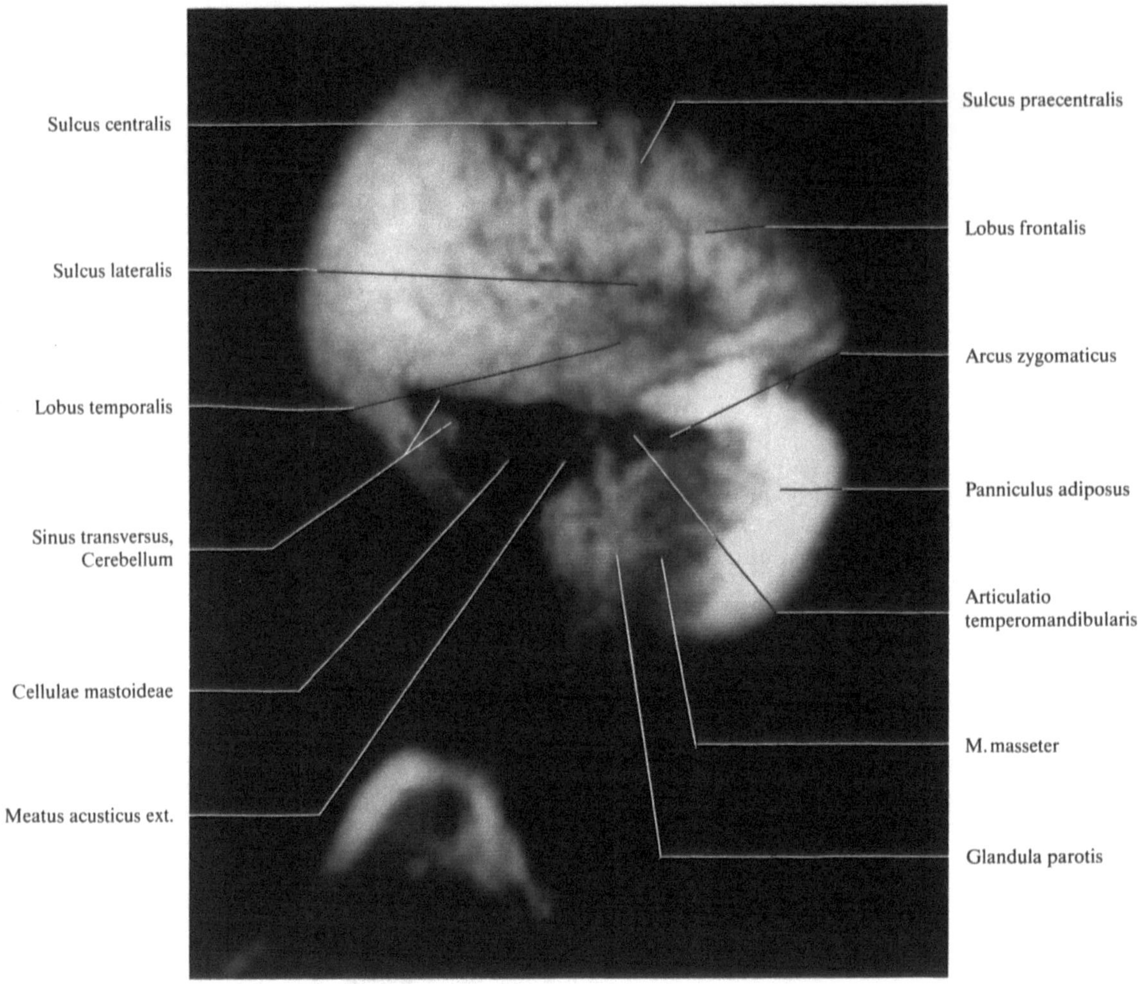

The central structure to be seen in section S7 is the temporomandibular joint (articulatio temporomandibularis), together with the zygomatic arch contiguous with it frontally and the external auditory meatus (meatus acousticus externus) located dorsally. Cranially the cellulae mastoideae cannot be distinguished from the parietal bone and the sinus transversus on the basis of differences in contrast. The brain is partly sectioned at the outermost margin, the sulcus lateralis roughly "bisects" the visualized cortex. As a result of its anatomical course the sulcus centralis is shifted frontally. The musculus masseter is quite well delimited, as a cranially directed area, from the panniculus adiposus and the parotid gland (glandula parotis). The subcutaneous vena jugularis externa can be distinguished faintly in the neck region.

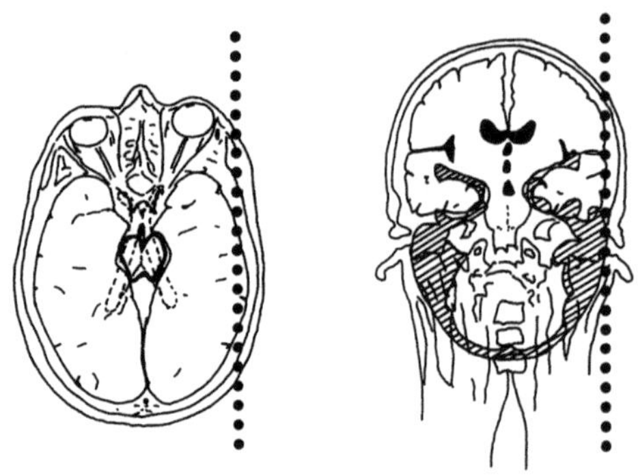

Sagittal section 6 cm lateral to the midline **S 7**

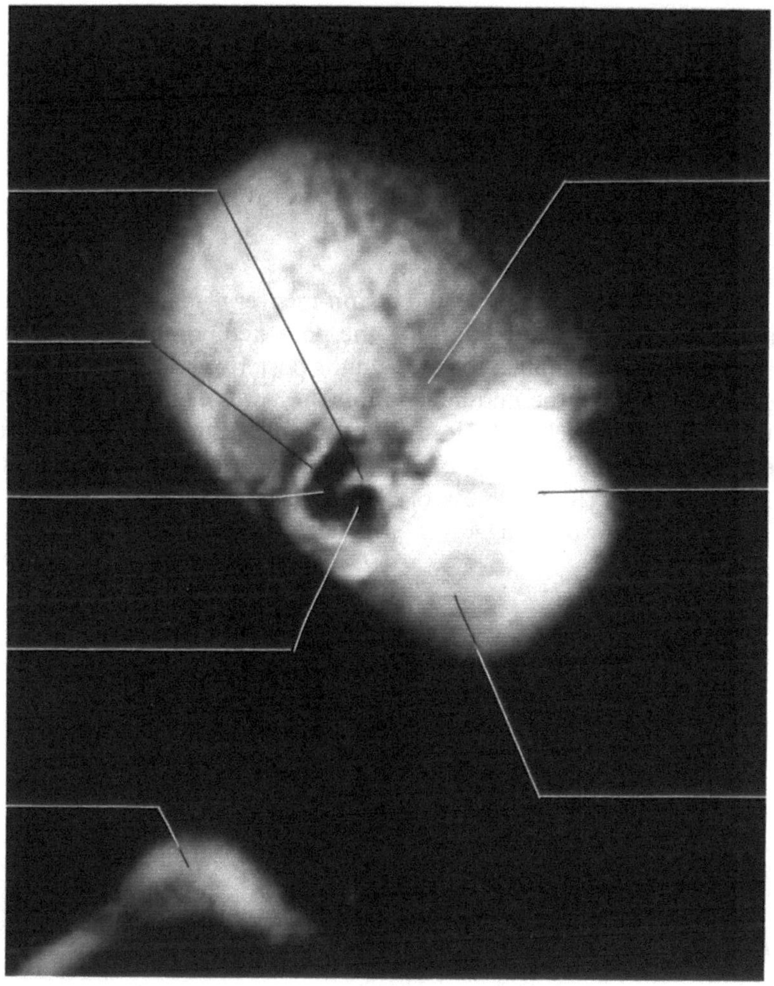

Section S8 is the outermost sagittal slice and only just touches the head, so that bone tissue and scalp overlap in the neurocranium, while in the facial region there is only a weak shadow of the musculus masseter and the clearly visualized panniculus adiposus. The delicate lines visible in the upper part of the skull are probably small blood vessels; thus, one line might represent the course of the arteria temporalis superficialis. Fine details are otherwise displayed in the external ear.

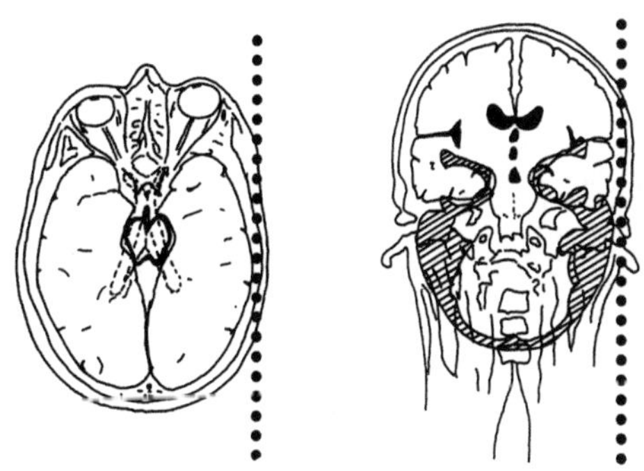

Sagittal section 7 cm lateral to the midline **S 8**

5. Frontal Sections

The first frontal section obtained was F1, the others followed at a spacing of 1 cm in the dorsal direction. F1 is approximately parallel to the plane of the face, passing through the middle of the eyeballs and the chin at the level of the foramen mentale. The final section in this series, F16, touches the confluens sinuum. The thickness of all slices was 12 mm. The corresponding inversion-recovery images begin frontally with F6 at the polus temporalis and end, dorsally, with F13 at the posterior horn of the lateral ventricles. Their thickness was 11 mm. Figures 5.1 and 5.2 present the middle sections of the two orthogonal sectional planes, S1 and H12. The position of all frontal sections is indicated.

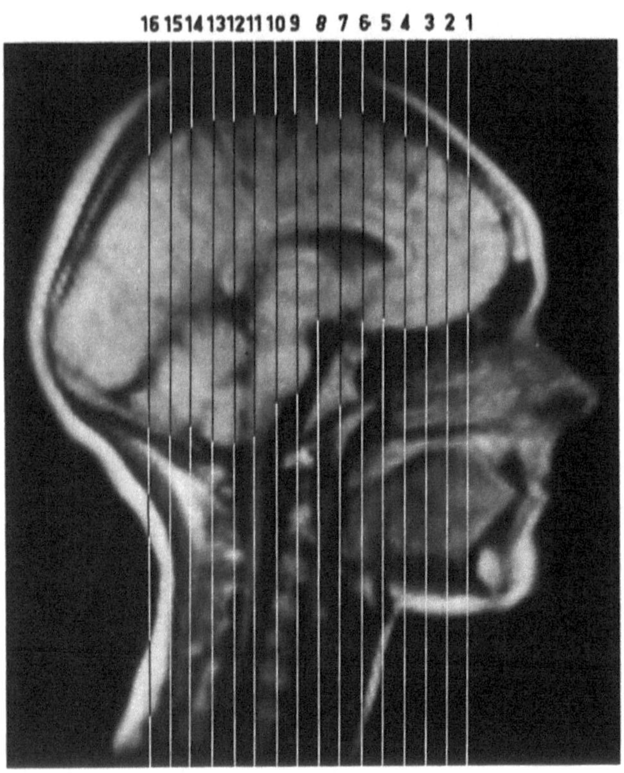

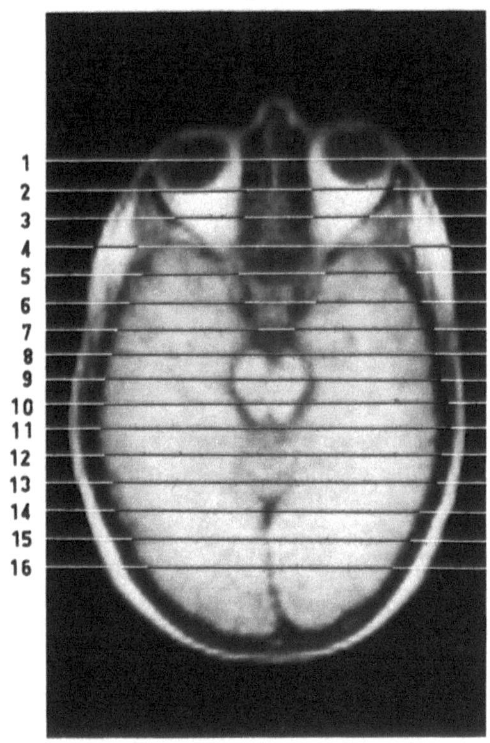

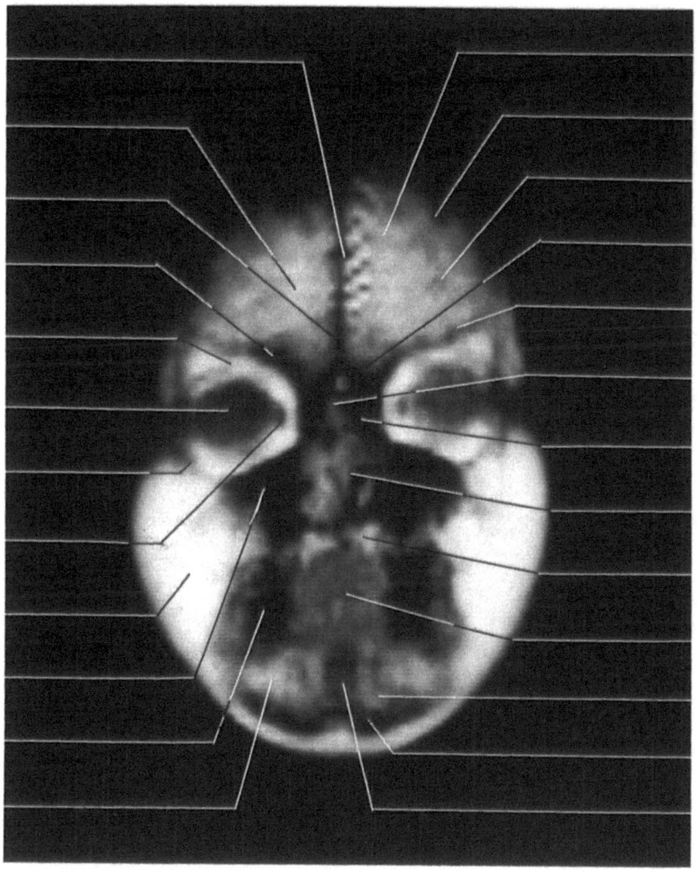

Section F1 passes through the eyeball slightly dorsally but is still ventral to the lateral boundary of the orbits. The sinus maxillaris appears at its fullest extent; the mandibula is visualized at about the level of the foramen mentale. The compact bone, as seen in the teeth, and the air-filled sinus produce few signals and thus appear black, while the subcutaneous fatty tissue of the face (panniculus adiposus) and the orbital fat appear very light by contrast. The lacrimal gland can be distinguished in the latero-cranial region of the orbit. The punctiform phenomena in the falx cerebri region of the frontal brain are artefacts. The enlargement of the right nasal conchae can also be recognized in the horizontal sections and is due to the fact that the subject was suffering from rhinitis.

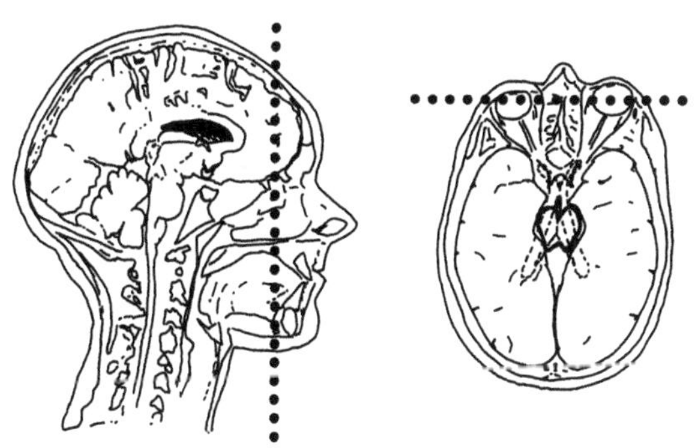

Frontal section 8 cm in front of the external acoustic meatus **F 1**

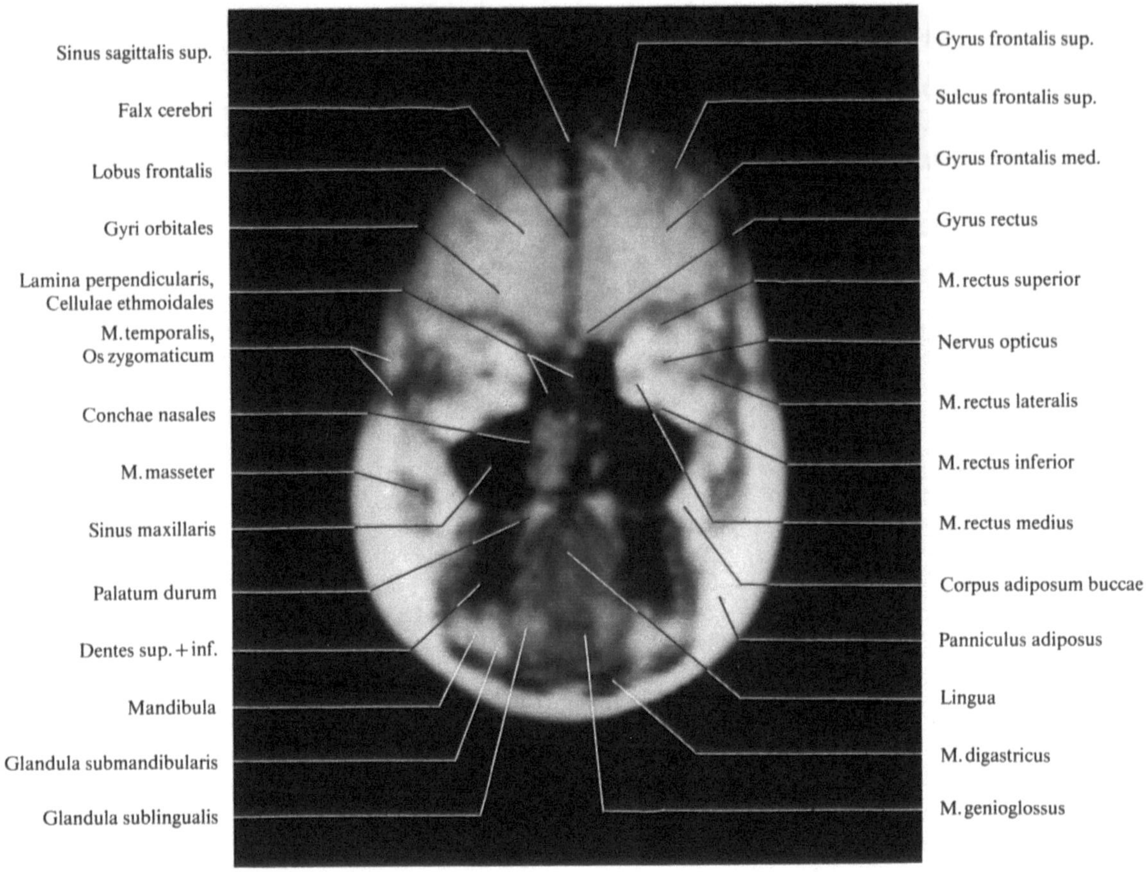

In section F2 the optic nerve, surrounded by the rectus muscles of the eye, is clearly visualized. In the frontal lobe the first convolutions (gyri) can be recognized; the light structures in the caudal prolongation of the falx cerebri are probably the filae olfactoriae. Between the corpus adiposum buccae and the panniculus adiposus the musculus masseter can be seen as a structure showing little contrast; the teeth are visualized as black areas above the mandible and the glandula submandibularis.

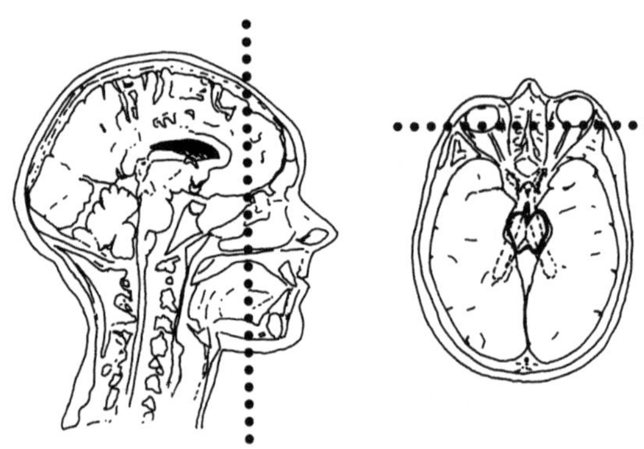

Frontal section 7 cm in front of the external acoustic meatus **F 2**

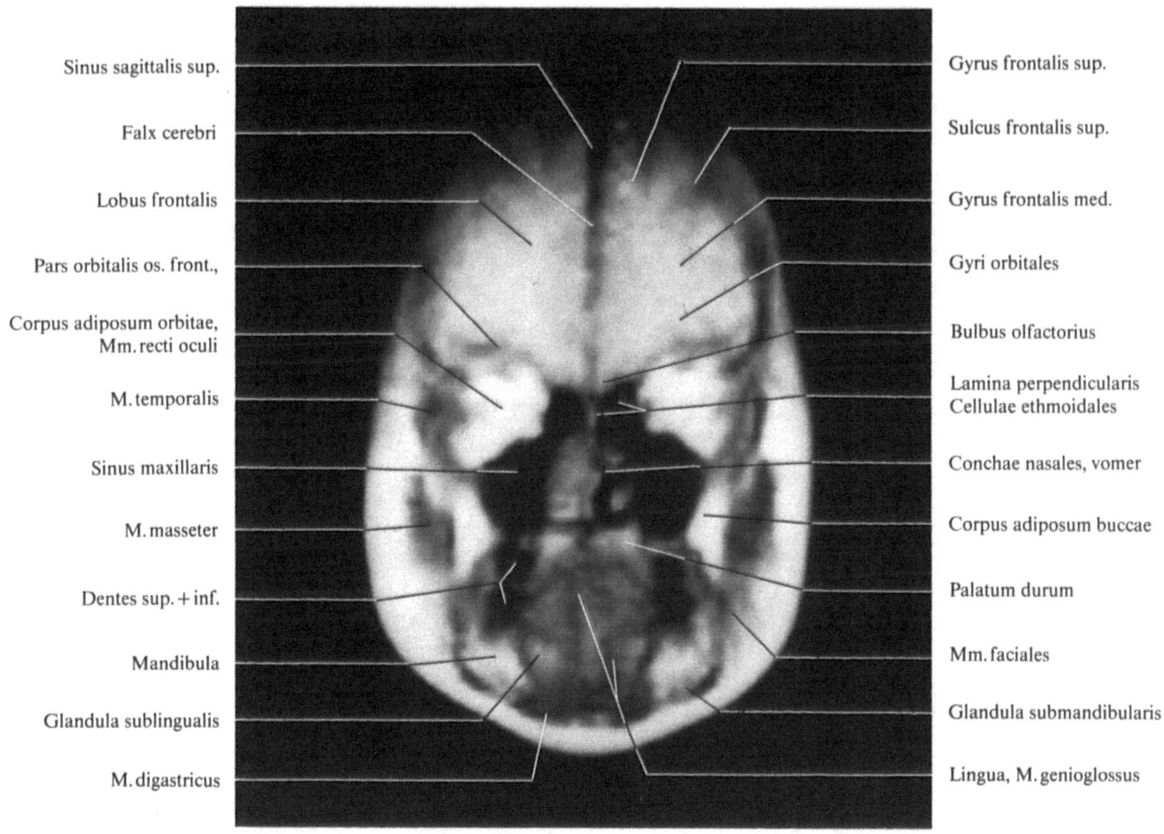

Section F3 reveals the dorsal third of the orbits; the adipose tissue is visualized as relatively homogeneous light areas, in which the muscles of the eye and the optic nerve cannot be differentiated. Laterally adjacent, the musculus temporalis appears as a signal-poor, inhomogeneous structure. The bony orbit is resolved only in the cranial portion, where it is sectioned approximately at right angles. Medially, no differentiation from the maxillary sinus is possible; laterally, the bone is too thin to provide any contrast at the given width of the section.

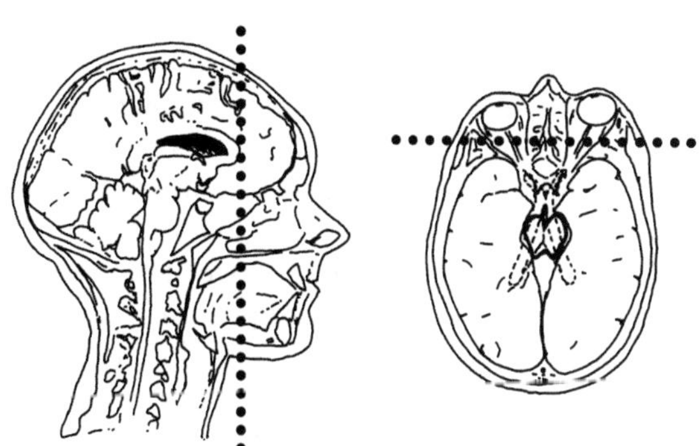

Frontal section 6 cm in front of the external acoustic meatus **F 3**

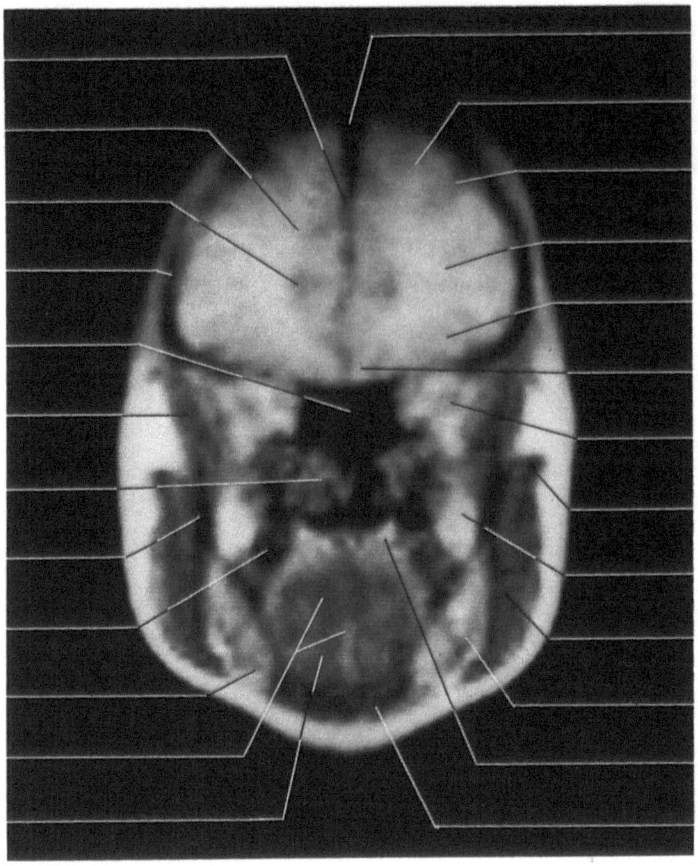

Section F4 passes through the anterior horns (cornu anterius) at their most ventral point and through the orbits almost at the level of the fissures. Here, the optic nerve and the tendons of the rectus muscles of the eye, which meet like the tip of a pyramid, are readily recognizable. The two frontal lobes are still completely separated by the dark falx cerebri. The mandible is visualized as a complex inhomogeneous structure in which both high-contrast bone marrow and low-contrast cortical bone can be recognized. Between the musculus masseter and the corpus adiposum buccae the processus coronoideus of the mandible is visible as a structure of purely compact bone. The teeth are still just represented by an upper molar.

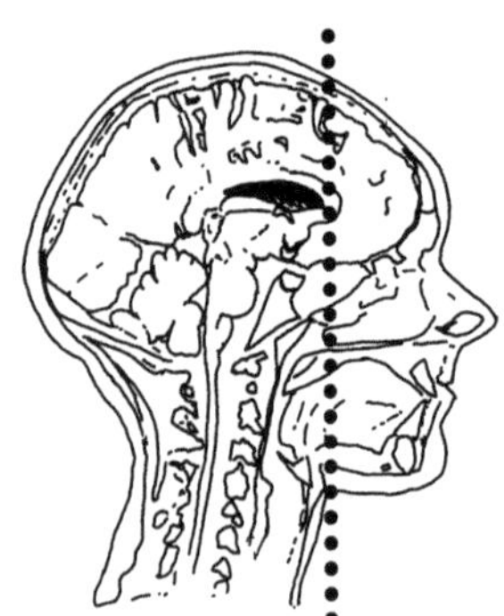

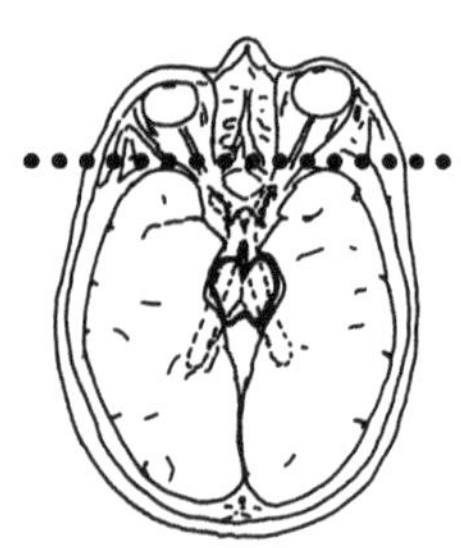

Frontal section 5 cm in front of the external acoustic meatus **F 4**

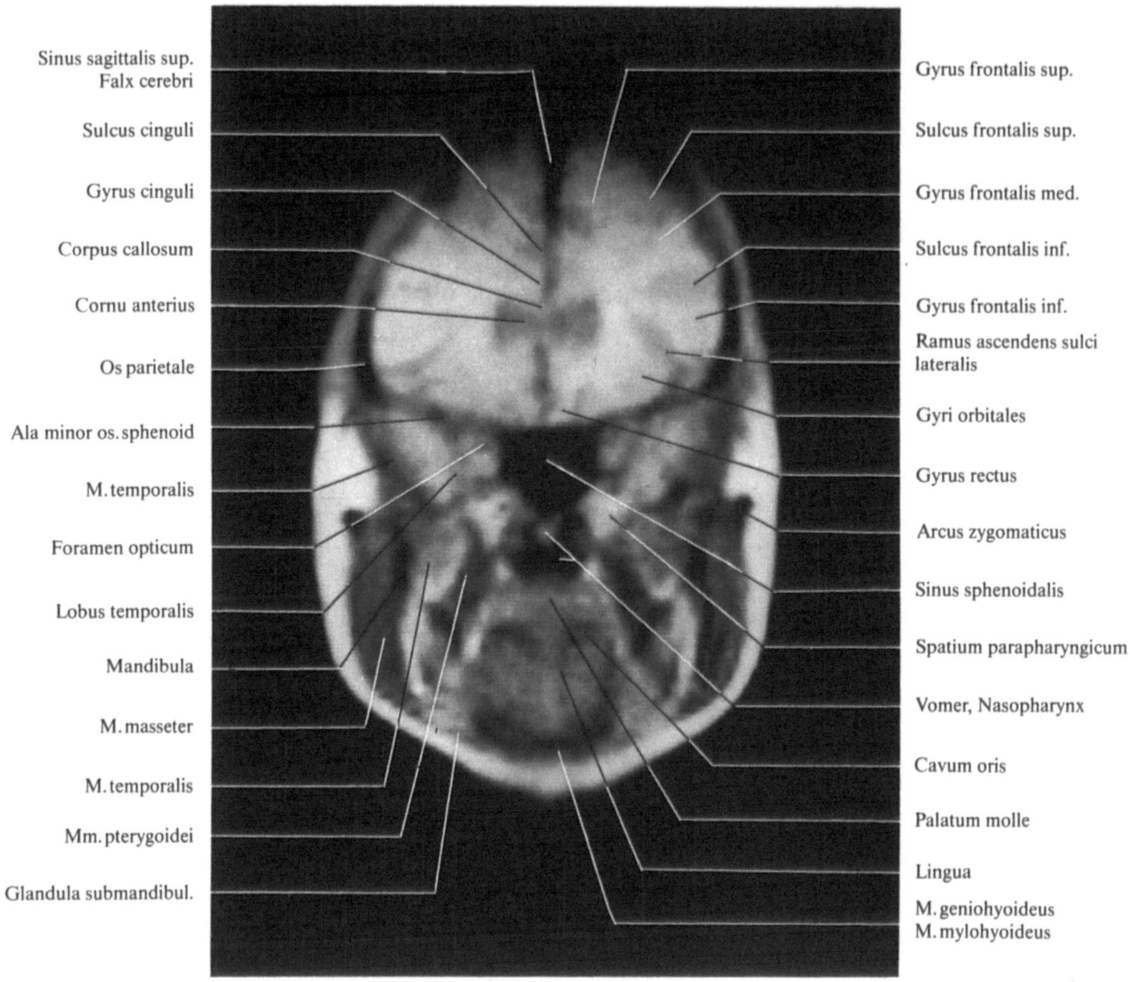

In section F5 the temporal lobes are visualized together with their poles. Medial orbit structures can still be recognized; they are separated from one another by the pneumatized, i.e. black, sinus of the sphenoid bone (sinus sphenoidalis). The lateral ventricles are sectioned at the level of the genu of the corpus callosum. The gyrus and sulcus of the corpus callosum and the cingulum are distinct. In the splanchniocranium the predominant structures are the large masticatory muscles, the musculus masseter and musculi pterygoidei, which form a frame around the ascending branch (ramus) of the mandible. The musculus temporalis medial to the zygomatic arch (arcus zygomaticus) extends upwards to the area of the temple. The nasal area is separated from the oral cavity by the soft palate.

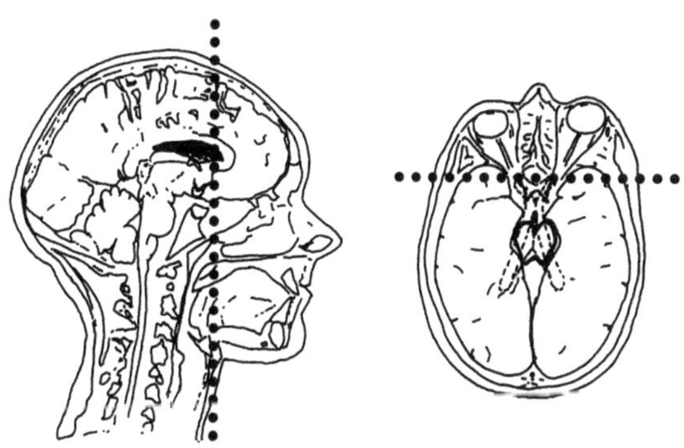

Frontal section 4 cm in front of the external acoustic meatus **F 5**

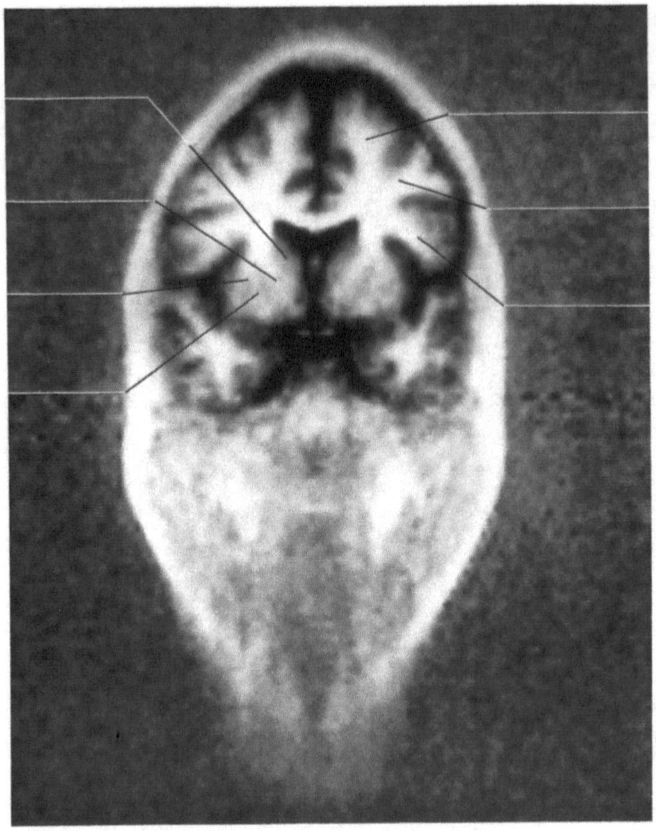

Fig. F6a

In section F6 the neurocranium predominates. The dark lateral ventricles join ventrally at the septum pellucidum. The frontal and temporal lobes, the diencephalon, the optic chiasma and the hypophysis are all sectioned. The nuclei and the tracts of the diencephalon are well resolved in the inversion-recovery scan F6a. In the caudal part of the section finer structures can be recognized only in the spin-echo image. Thus, the ascending branch of the mandible, differentiated into cortical (dark) and cancellous (light) bone and framed on both sides by the large masticatory muscles, can readily be recognized. The nasopharynx and the pharynx join at the uvula; in the caudal part of the image the hyoid bone (os hyoideum) can be seen.

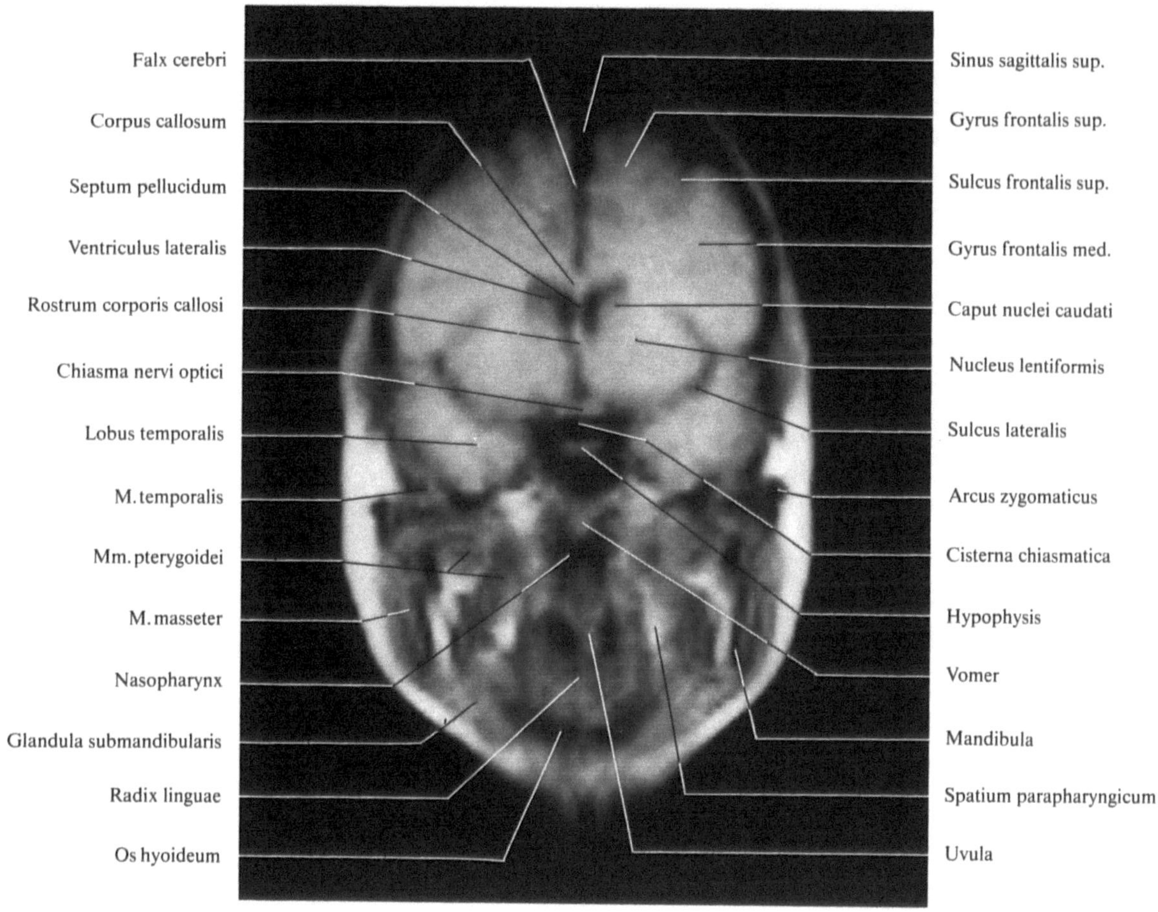

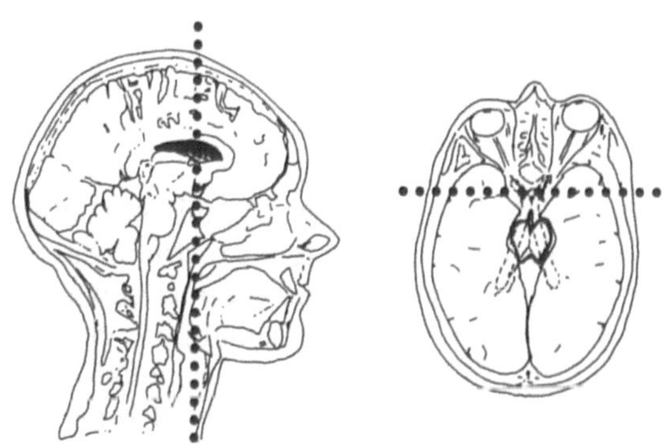

Frontal section 3 cm in front of the external acoustic meatus **F 6**

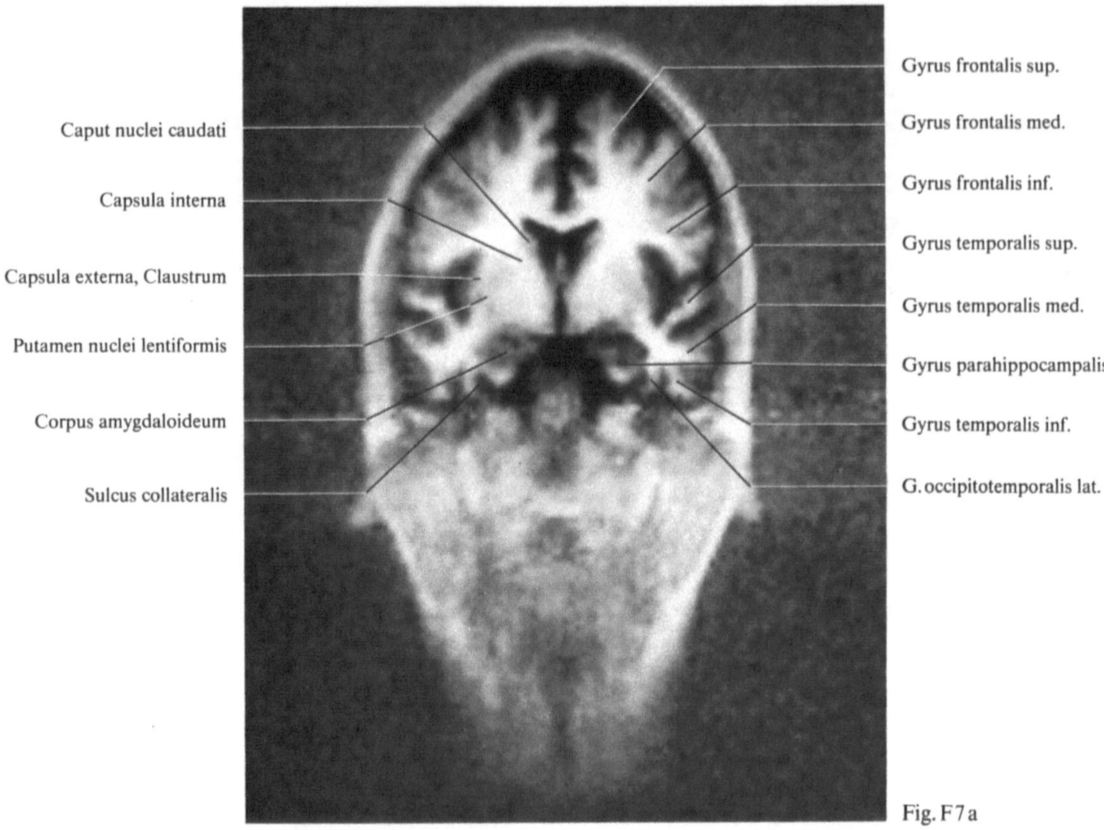

Fig. F7a

Section F7 passes through the commissura anterior to the epiglottis. Caudal to the rostrum and the third ventricle the chiasma nervi optici, separated from the hypophysis by the signal-poor cisterna chiasmatica, can be recognized. Convolutions of the frontal lobe, insula and temporal brain are again clearly visualized in the inversion-recovery image F7a. Here, diencephalon structures, such as the nucleus lentiformis, the caput nuclei caudati and the internal and external capsules, are also distinguishable. In the region of the hippocampus the corpus amygdaloideum can be recognized. Lateral to the mandible, the glandula parotis can be seen forming the dorsal continuation of the musculus masseter; caudally, the glandula submandibularis is visualized; both contrast markedly with the subcutaneous adipose tissue and the adjacent muscles.

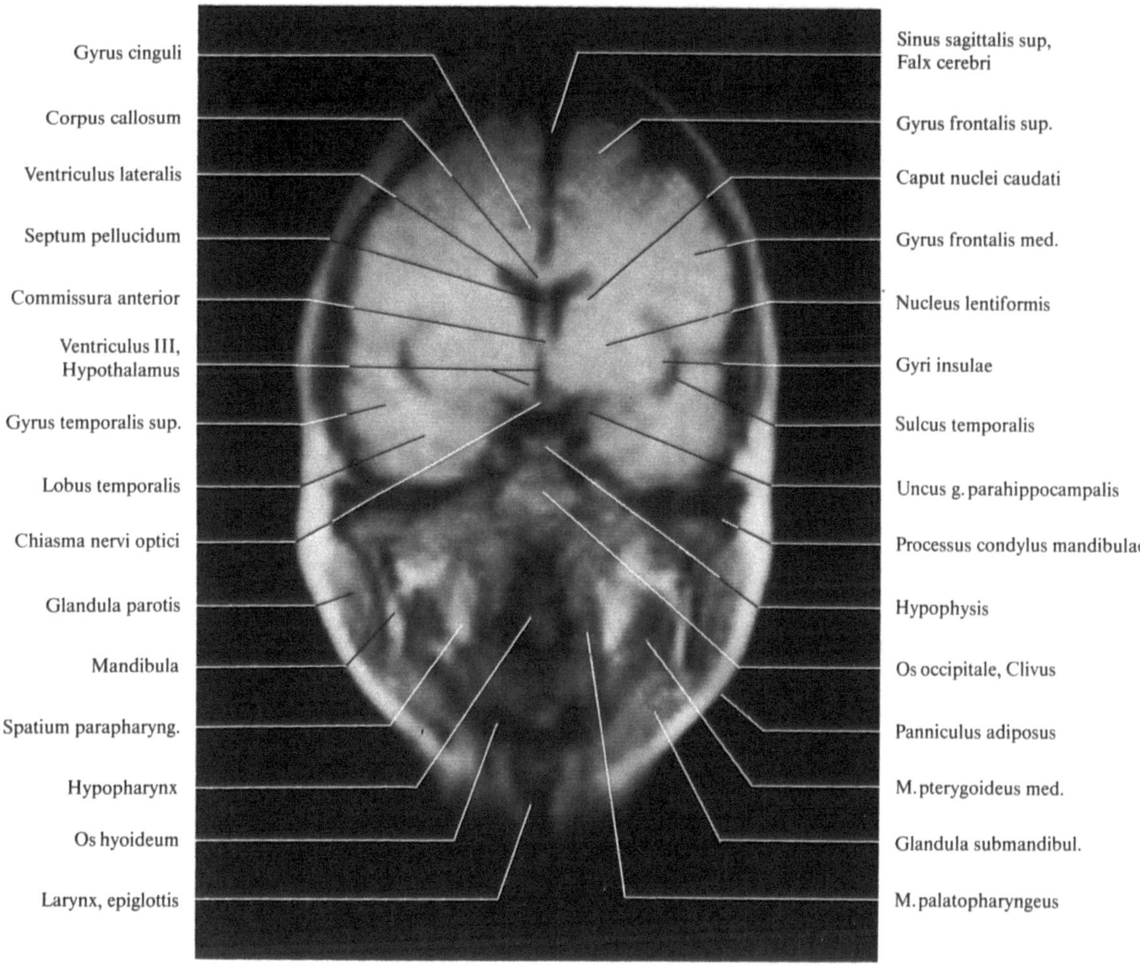

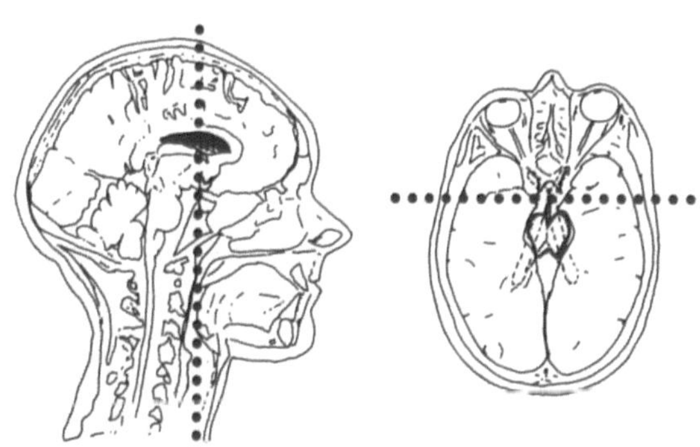

Frontal section 2 cm in front of the external acoustic meatus **F 7**

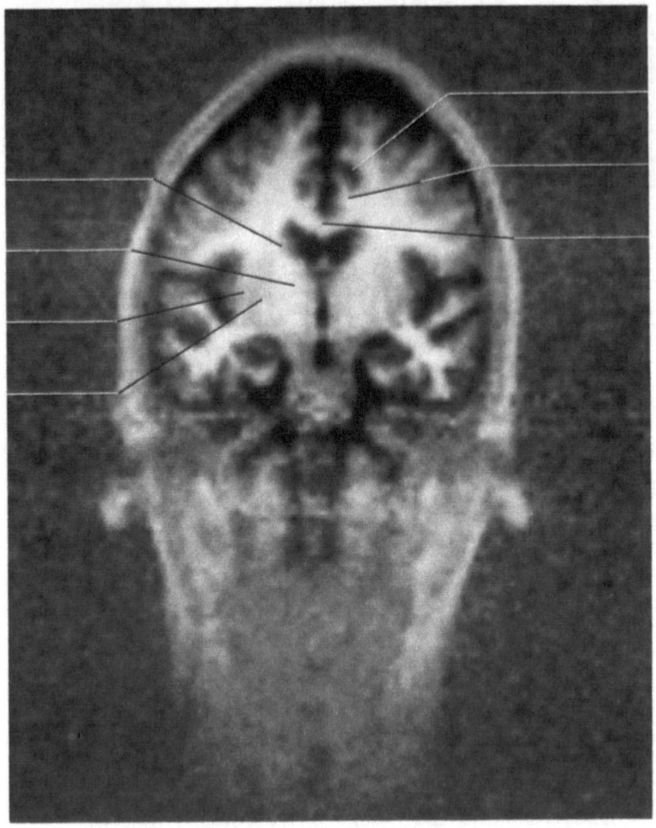

Fig. F8a

The plane of section F8 touches the ventral margin of the pons. There is a mixing of the signals obtained from the cisternae and tissue of the midbrain, such that the latter is visualized darker than normal. The frontal lobe is sectioned at the junction with the lobulus paracentralis; the sagittal course of the sulci changes to a lateral one, as can be recognized particularly well in the inversion-recovery image F8a. The nuclei of the diencephalon, with the exception of the nucleus caudatus, are very indistinctly visualized in this latter image. Located laterally, above the glandula parotis, the temporomandibular joint can be seen. In the neck the larynx has been sectioned at its narrowest part; above the larynx the recessus piriformis and, more laterally, the great vessels of the neck are to be seen.

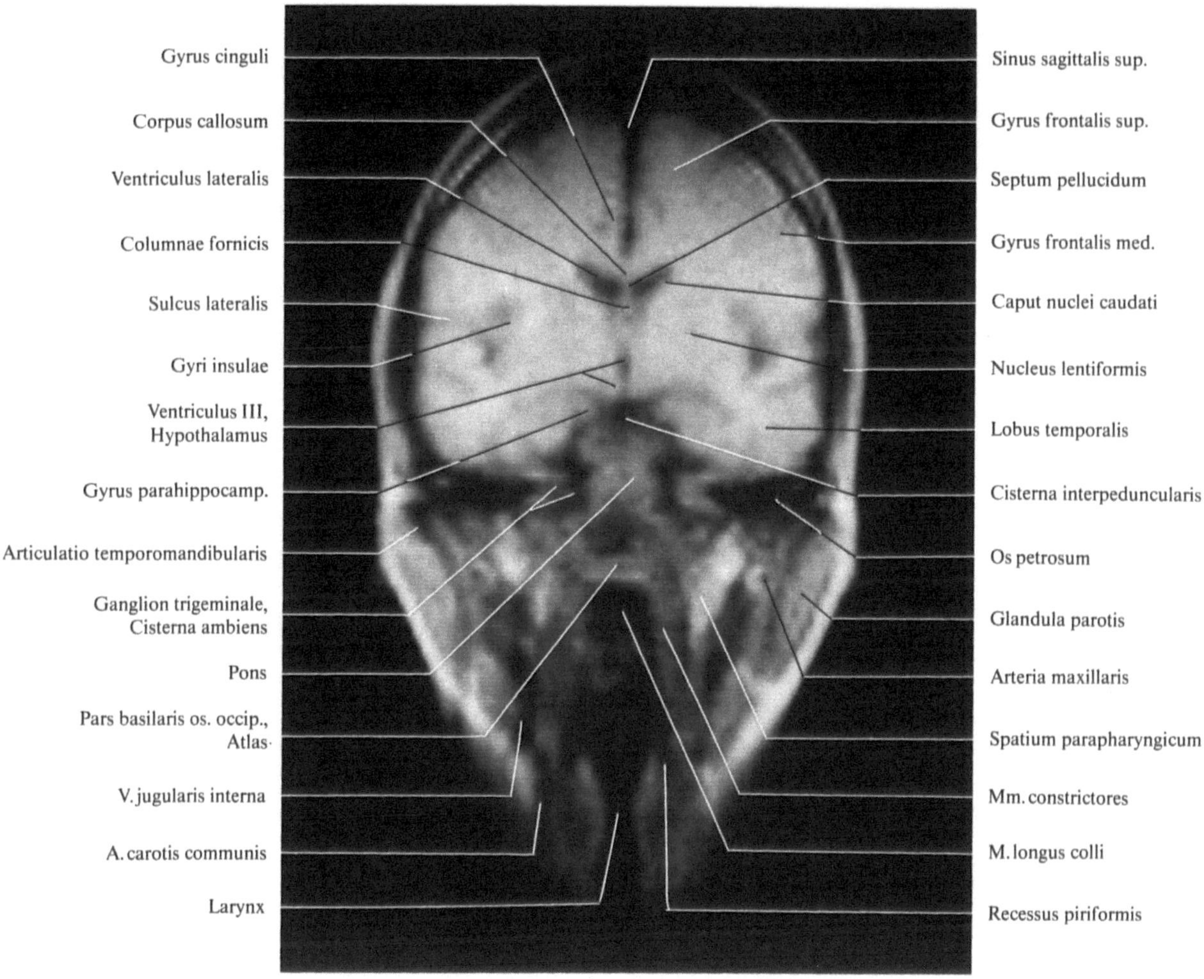

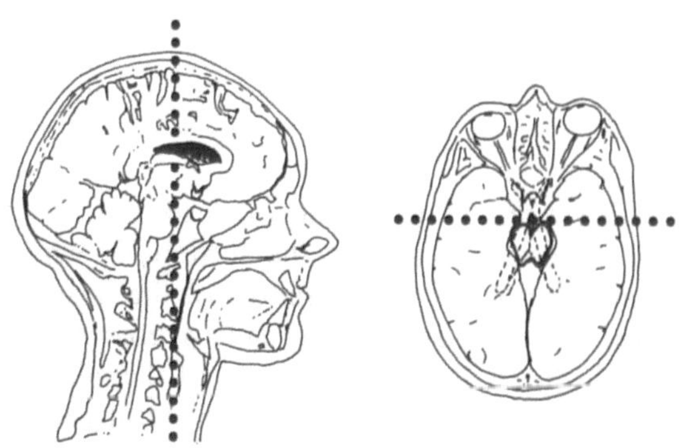

Frontal section 1 cm in front of the external acoustic meatus

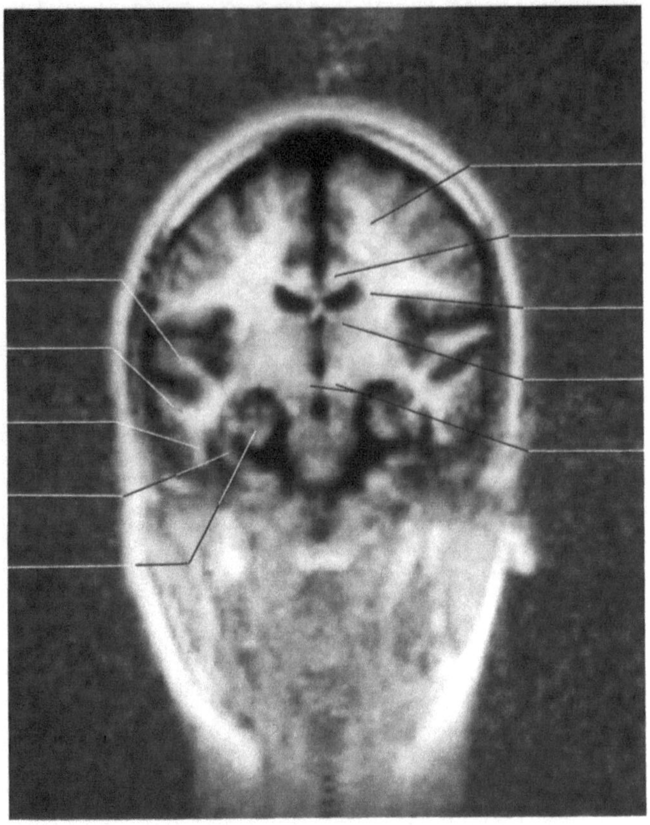

Fig. F9 a

The directions of the orthogonal sections have been indicated relative to the section F9, which lies approximately in the middle of the head. It passes from the area of the sulcus precentralis through the thalamus at the level of the adhesio interthalamica and the pons to the dorsal part of the larynx. Lateral to the pons the ganglion trigeminale is intensively visualized with, below, the occipital condyles and the basilar part of the occipital bone that forms the ventral limit of the foramen magnum. In the inversion-recovery image F9 a the diencephalon reveals hardly any structures; only the convolutions of the temporal lobe can be readily identified, while the section runs roughly parallel to the gyrus precentralis, so that this latter presents a very unstructured appearance.

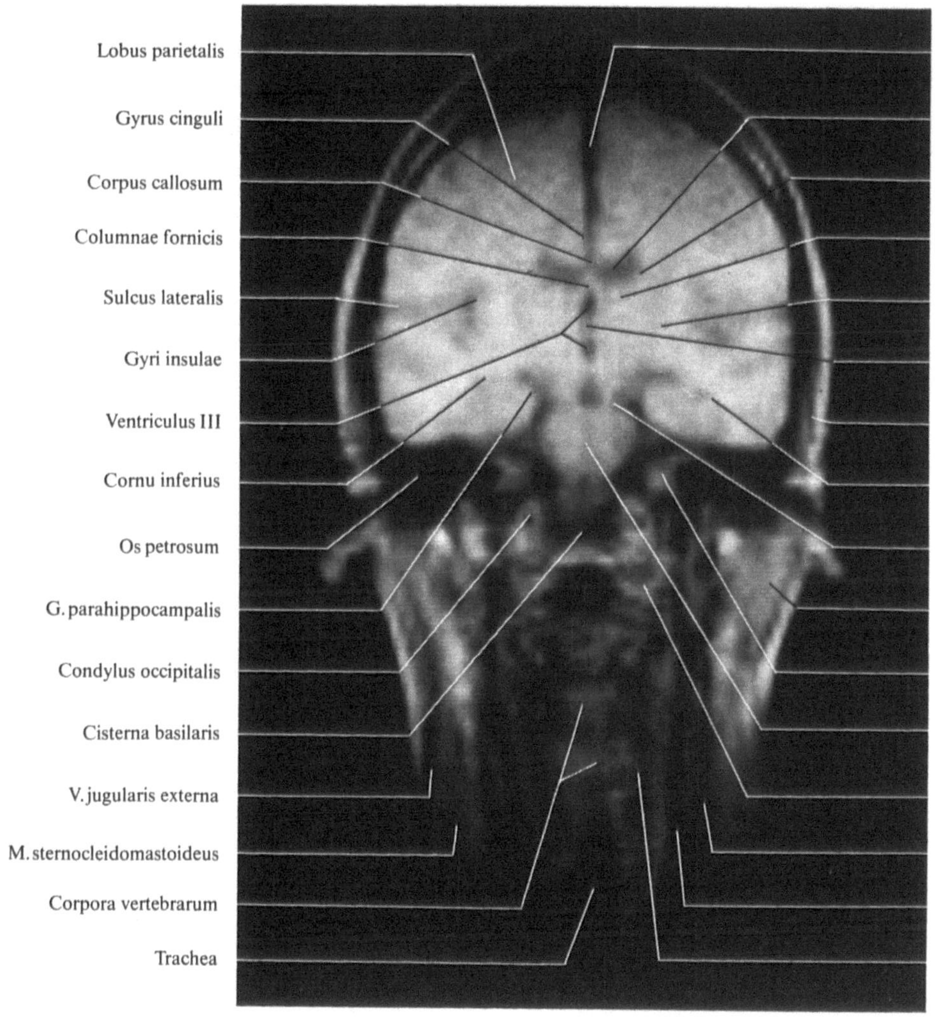

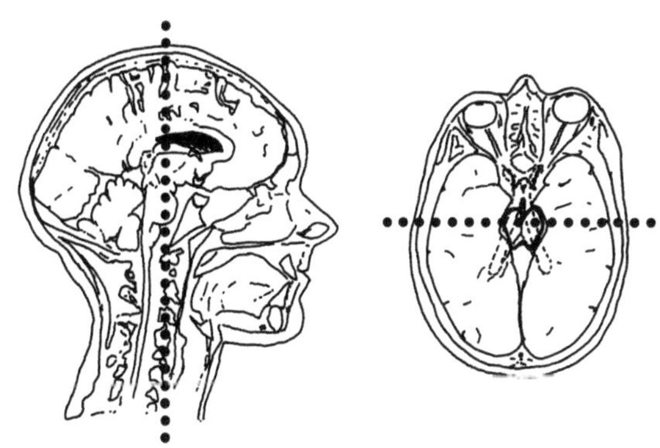

Frontal section through the external acoustic meatus **F 9**

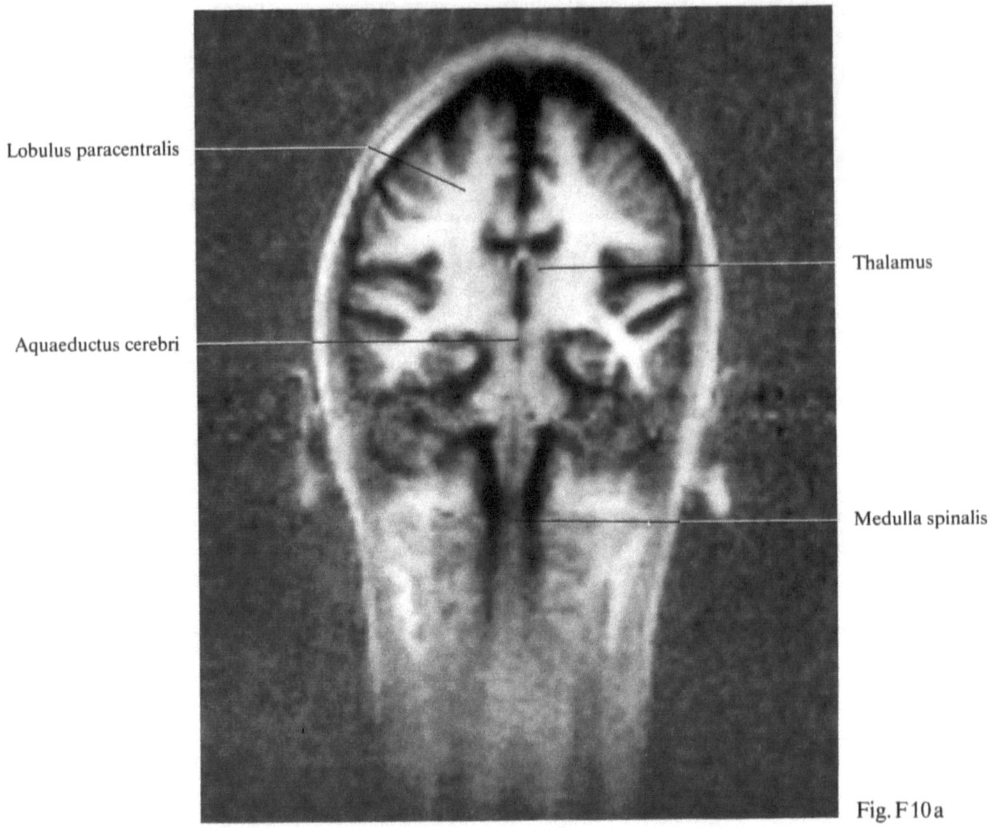

Fig. F10a

Section F10 shows the posterior portion of the third ventricle and the aqueductus cerebri; it passes through the tectum, the pedunculi cerebelli and the medulla oblongata. In the neck region the vertebral bodies are visualized with the surrounding ligamentous structures and the major muscles. The inversion-recovery image F10a is slightly tilted with respect to the spin-echo image; the cisterna ambiens cannot yet be seen, while the spinal cord can be followed further caudally.

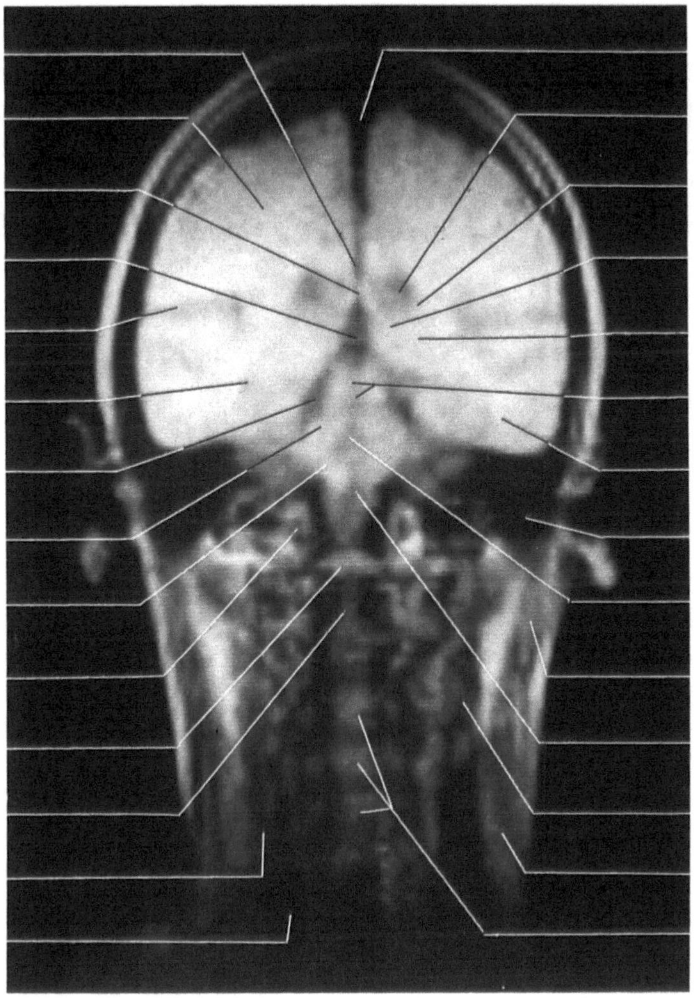

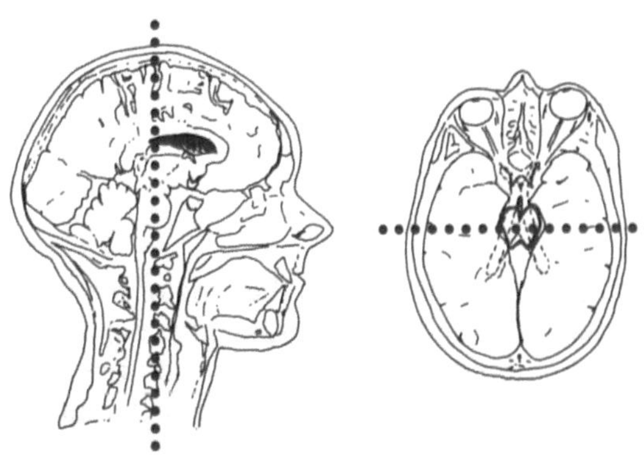

Frontal section 1 cm behind the external acoustic meatus

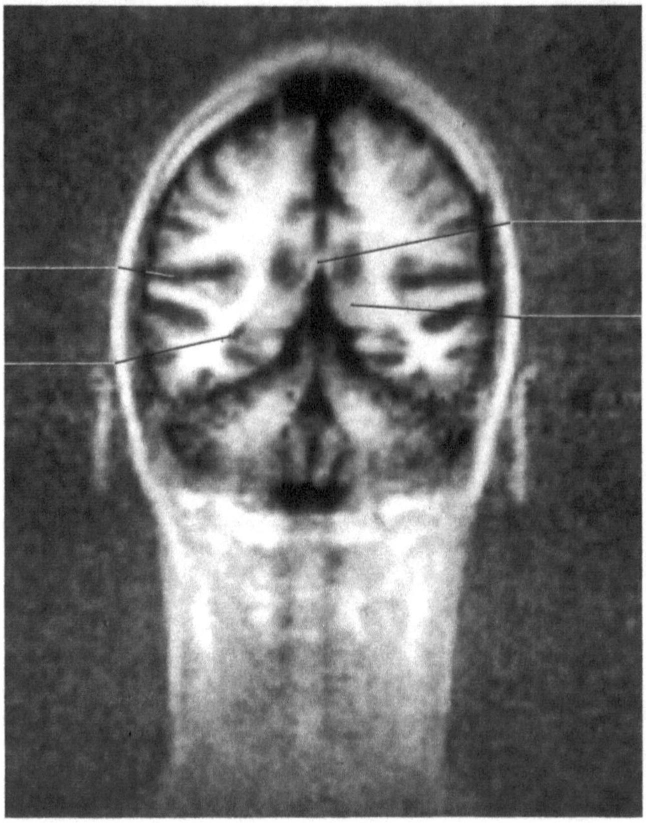

Fig. F11a

Section F11 passes through the atria of the lateral ventricles, the epiphysis (corpus pineale) and the fossa rhomboidea, which merges with the spinal canal. The spinal cord is displayed down into the lower parts of the neck, but with interference from structures of the vertebral canal. In the inversion-recovery image F12a the white matter of the corpus callosum can be seen between the lateral ventricles and, in the left-hand part of the image, the narrow, dark cornu inferior (inferior horn). Both in the cerebellum and in the cerebrum marked differences in contrast between the tracts comprising the white matter and the cells and synapses comprising the grey matter are evident.

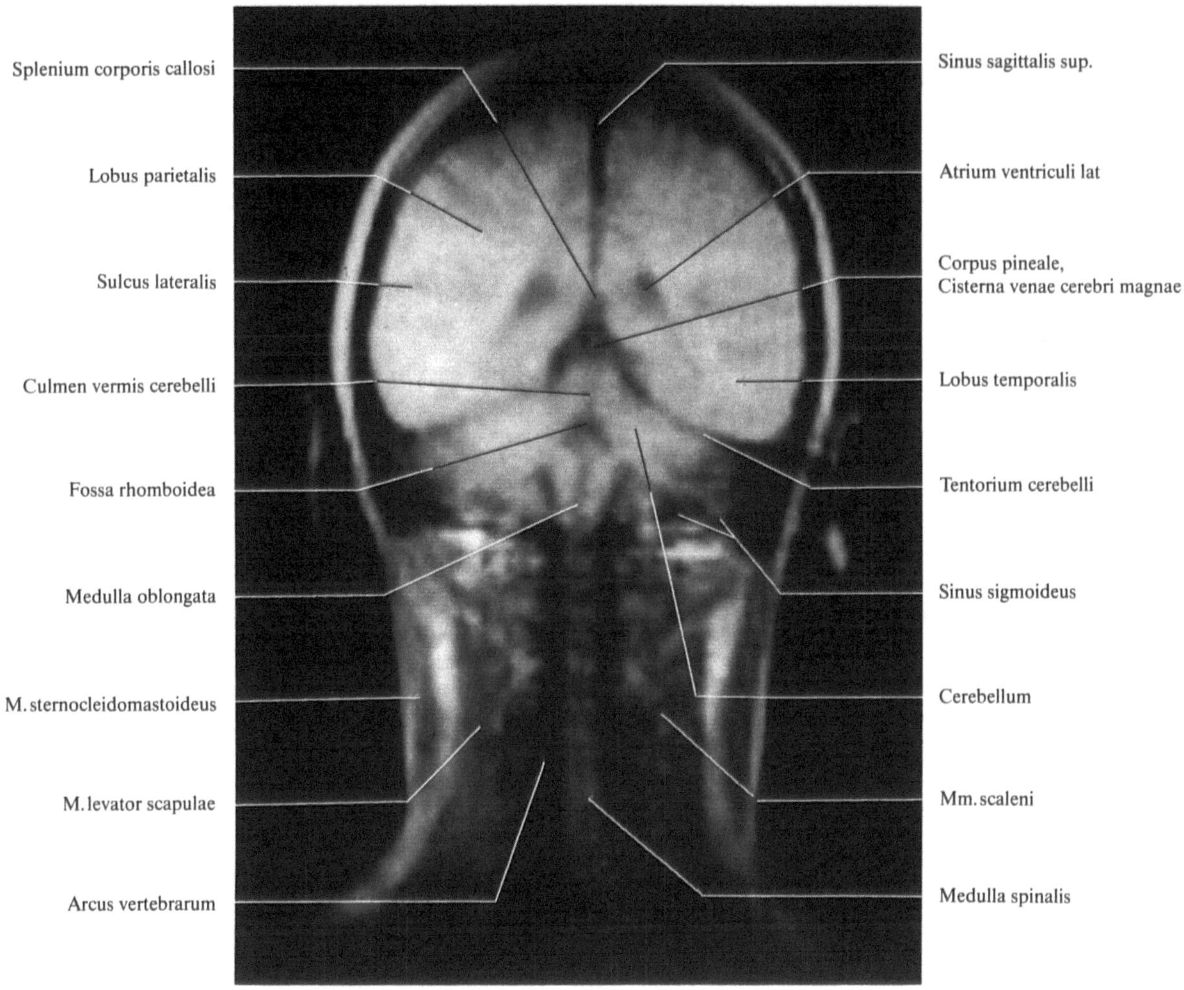

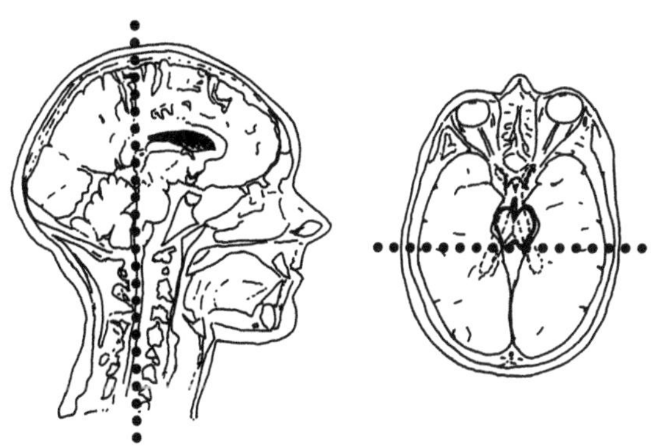

Frontal section 2 cm behind the external acoustic meatus **F 11**

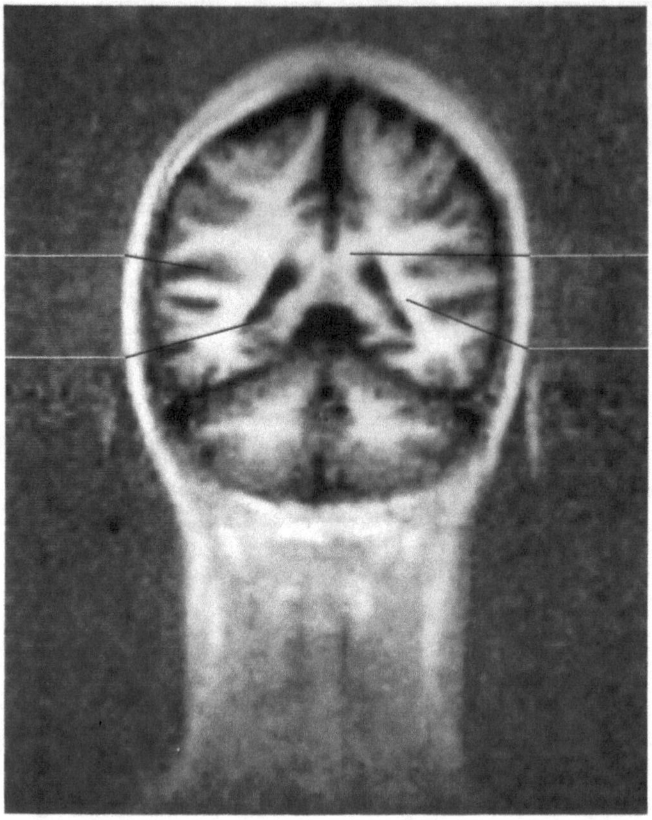

Fig. F12a

Section F12 cuts the lobulus paracentralis in the region of the sulcus postcentralis, the splenium of the corpus callosum and the cerebellum with the tonsil. In the inversion-recovery scan F12a the parietal and temporal lobes, separated by the sulcus lateralis but merging to an ever greater extent, can be recognized. In the neck, structures of the vertebral column and its ligaments are visualized. With respect to the muscles, the musculus sternocleidomastoideus is extremely prominent in the lateral position and can also be recognized very well in the first horizontal sections.

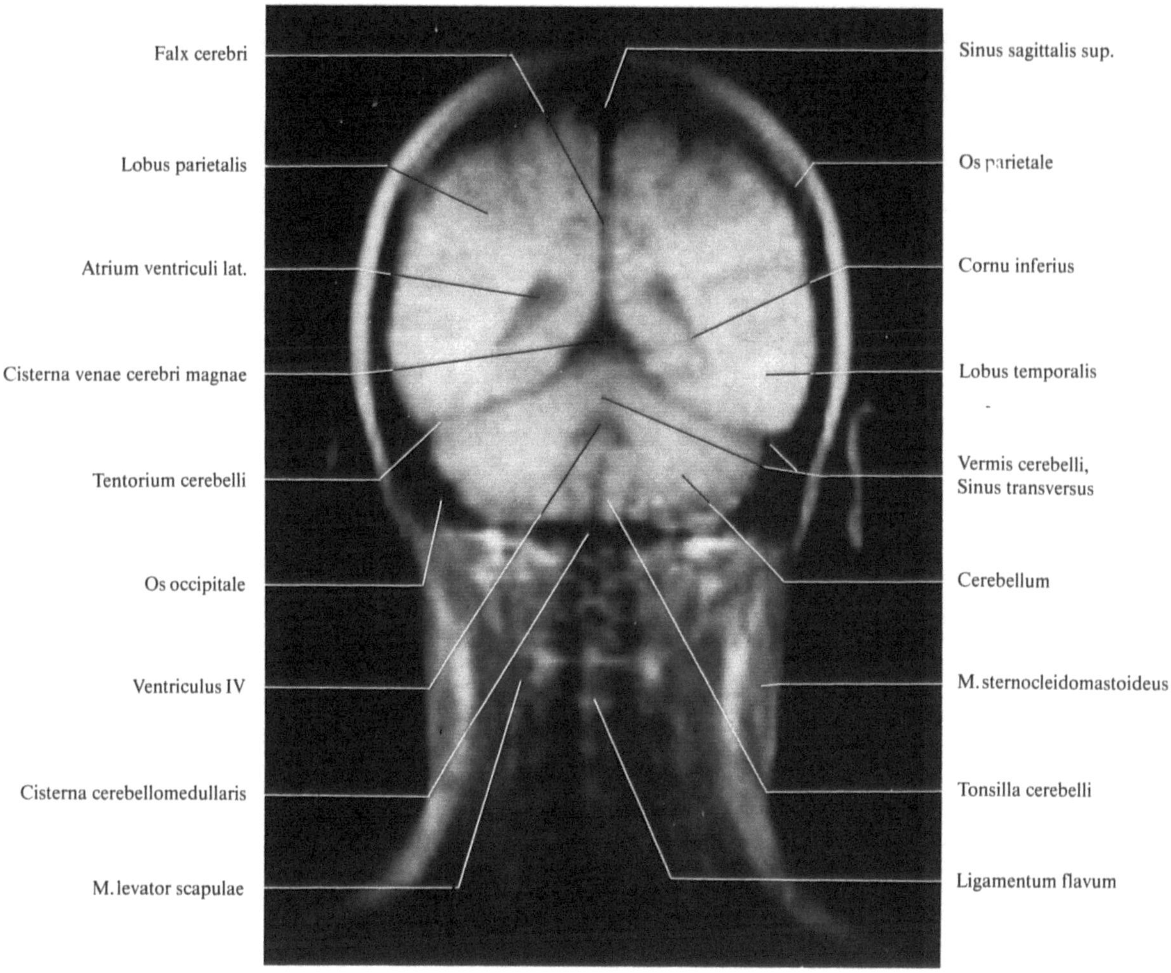

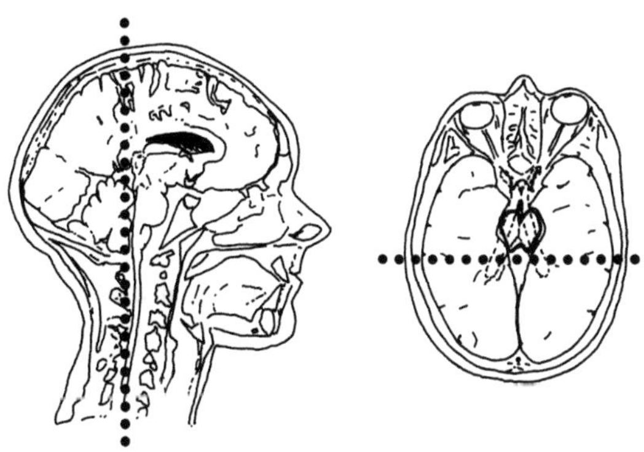

Frontal section 3 cm behind the external acoustic meatus **F 12**

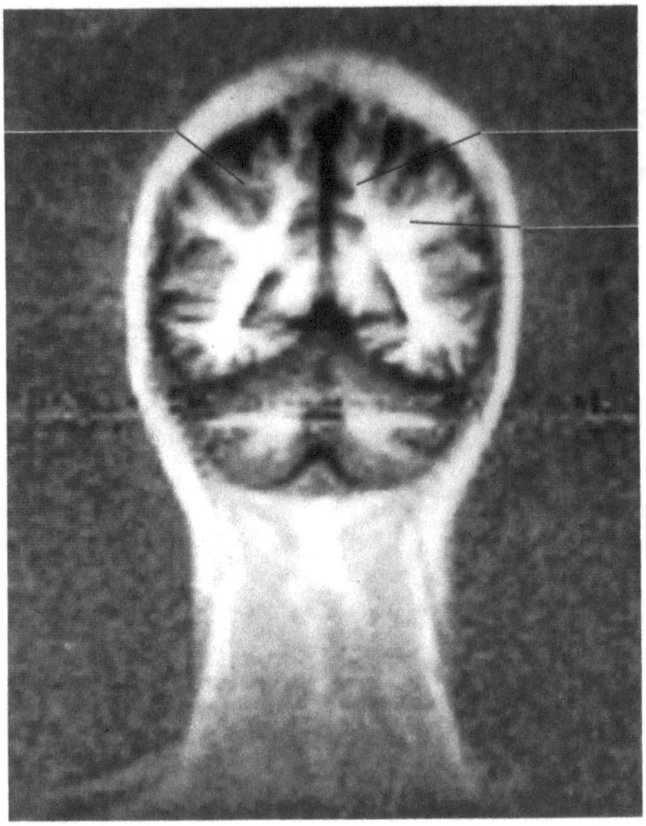

Fig. F13a

In section F13 the sulcus intraparietalis, which separates the lobulus parietalis superior and inferior, is seen. As in the frontal brain the structures take a sagittal course, and are thus well visualized. The posterior horns of the lateral ventricles are now narrower, and their negative contrast is superimposed by more powerful signals generated in the brain tissue. The falx completely separates the two hemispheres and expands caudally to form the sinus rectus and cranially to form the sinus sagittalis superior. In the neck, the ligamenta nuchae appears as a light triangular structure; caudal to this the spinous processes of the cervical vertebrae, together with the surrounding ligaments, which generate a major part of the signal, are visualized.

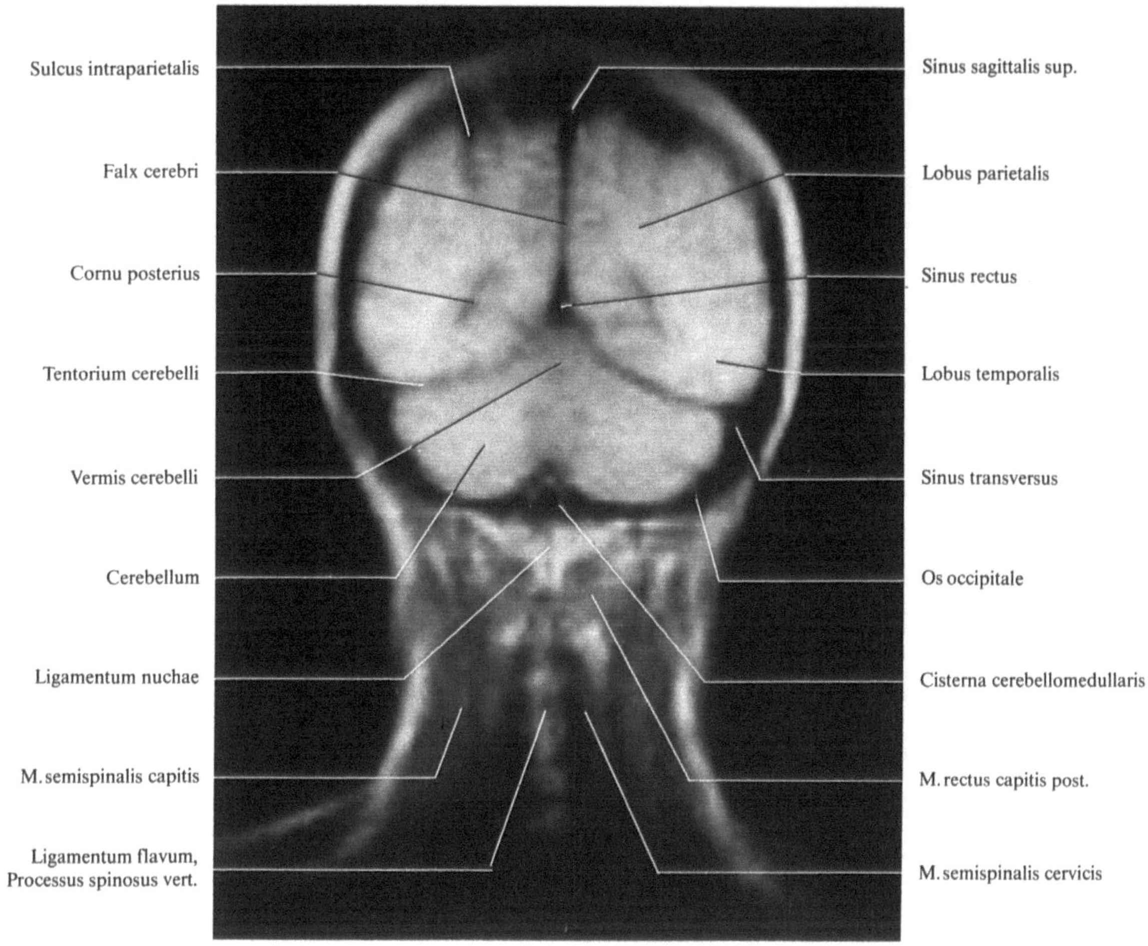

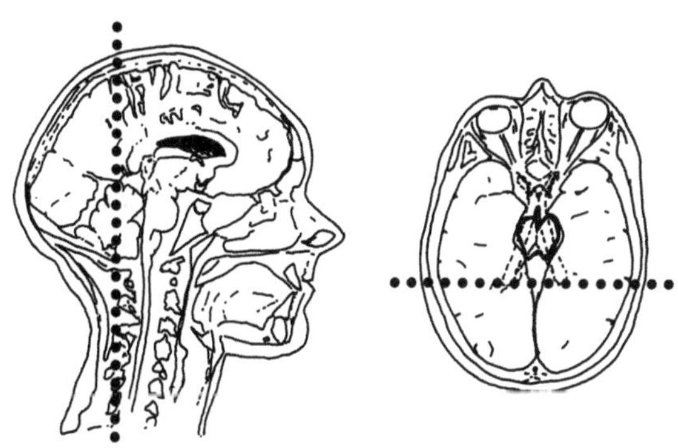

Frontal section 4 cm behind the external acoustic meatus **F 13**

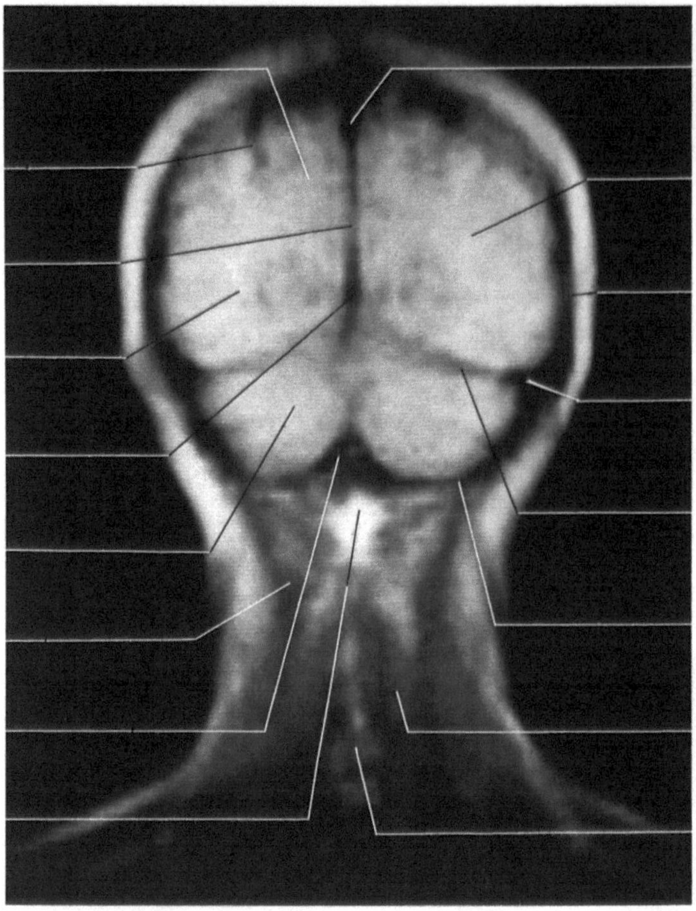

Section F14 passes exclusively through the parietal lobe and the cerebellum. The sulcus intraparietalis, which follows an approximately sagittal direction, clearly separates the two lobuli parietales. On a level with the designation "lobus parietalis", the gyrus angularis, the transition to the temporal lobe, is just visualized. The dark sinus rectus contrasts with the surrounding brain tissue. A bilateral diminishment in the intensity of the signal within the lobulus parietalis inferior, on a level with the sinus rectus, is the last sign of the posterior horns of the lateral ventricles. The sinus transversus can be recognized as a dark indentation into the tentorium cerebelli; in contrast, the tentorium here presents a lighter than normal appearance, since, owing to its oblique course with respect to the section plane, spatial integration occurs with the neighbouring tissues, which are richer in signals. Caudal to the large cisternae (cisterna cerebellomedullaris) and the occipital bone (os occipitale), which are both visualized as dark areas, the signal-rich ligamenta nuchae can be distinguished. The spinous processes (processus spinosus) appear as light structures compared with the bone marrow and the surrounding ligaments, which are strongly visualized.

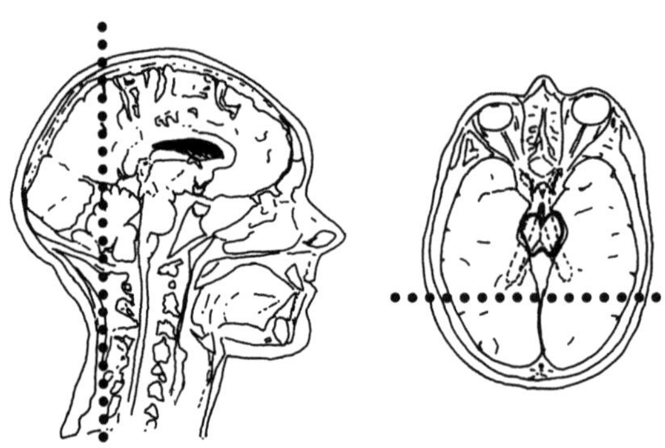

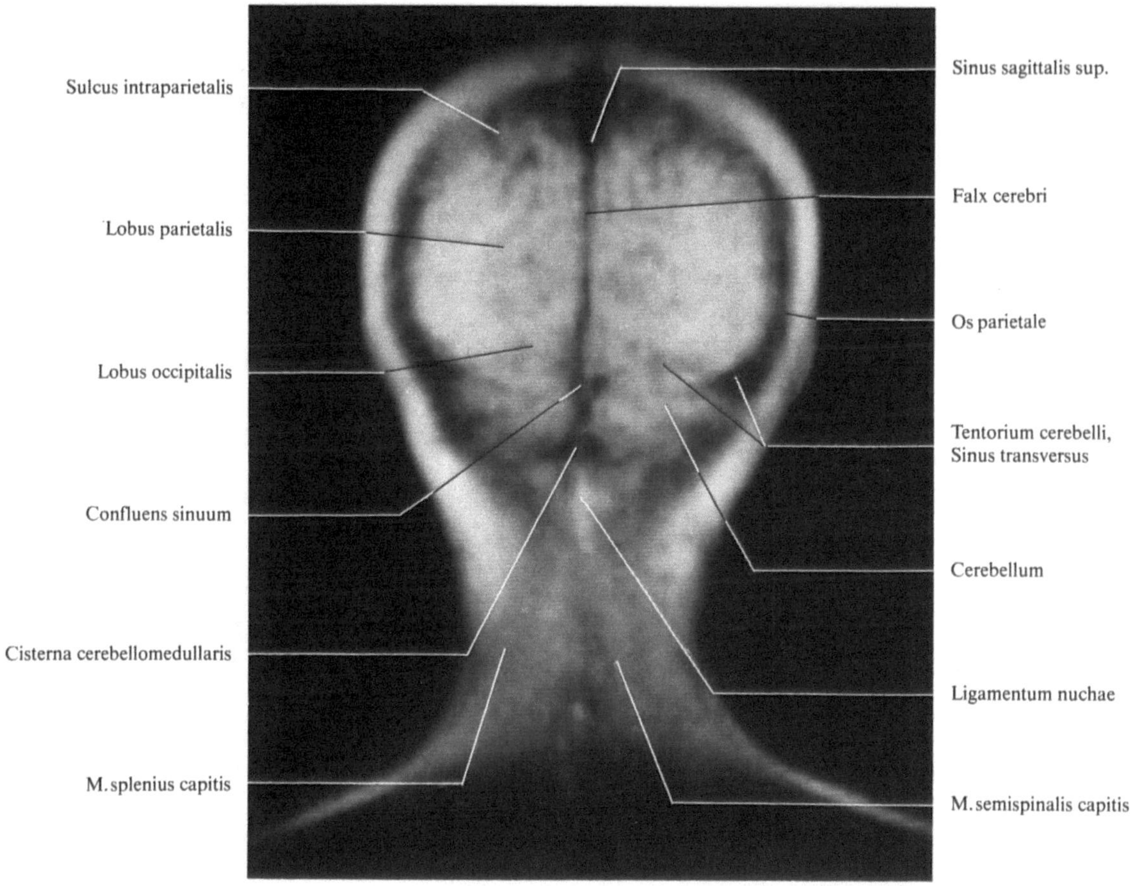

Section F15 cuts the dorsal boundary of the cerebellum and just touches the posterior lobes of the cerebrum; thus, the confluens sinuum is visualized. The sinus transversus can be seen lateral to the structures of the cerebellum; the tentorium is now hardly differentiable. In the parietal lobes, individual convolutions can be recognized, which, however, with the exception of the sulcus intraparietalis cannot be accurately identified. The muscles of the neck displayed in the image probably merge with the musculus trapezius in the middle region of the neck.

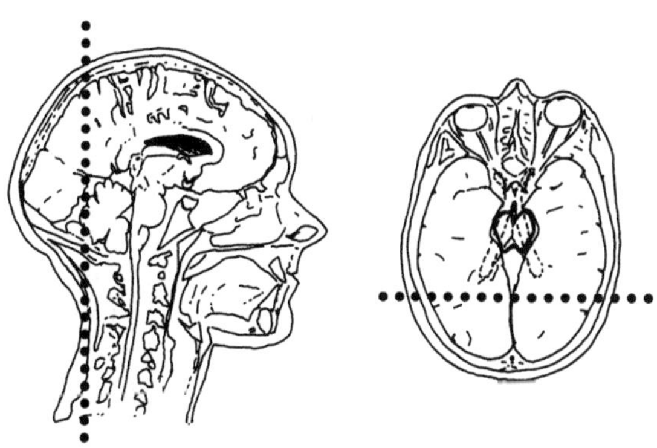

Frontal section 6 cm behind the external acoustic meatus **F 15**

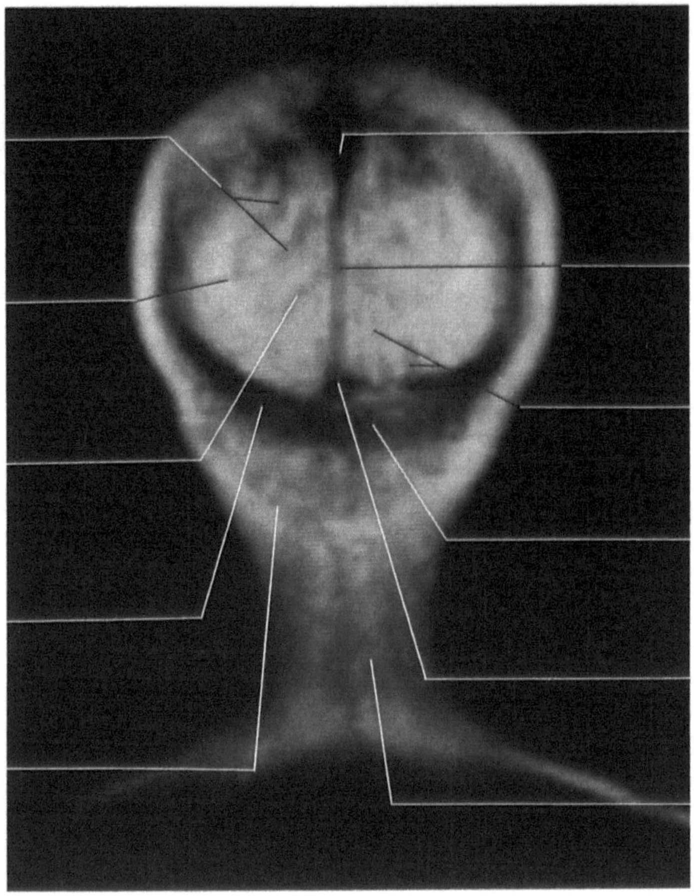

Section F16 shows the parietal and occipital lobes fairly equally; the line of separation between them is the sulcus parieto-occipitalis. The sulcus calcarinus arches over, in typical manner, the gyrus occipito-temporalis medialis, which, laterally, is in turn separated from the gyrus occipito-temporalis lateralis by the sulcus collateralis. In the neck the musculus trapezius can now clearly be identified.

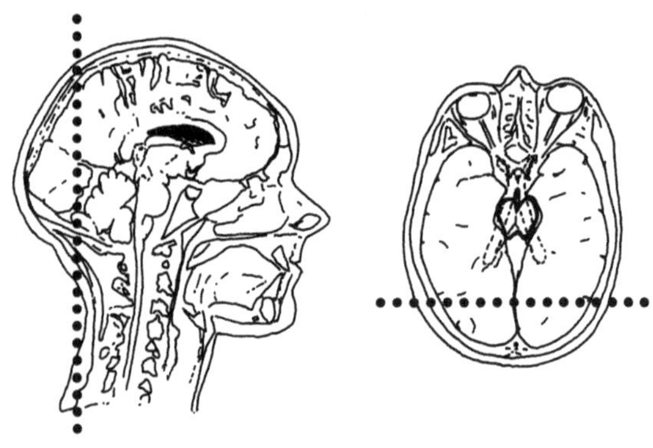

Frontal section 7 cm behind the external acoustic meatus **F 16**

6. Horizontal Sections

In contrast to the previous sections it proved expedient here to change the position of the subject twice while obtaining the 19 horizontal sections. In order to check the direction of the sectioning, which was intended to lie parallel to the plane of the optic nerve, the series started with section H12. Additional sagittal planes were measured prior to and after the change in position to ensure accuracy of the positioning. Section H1 lies approximately at the level of the chin and is followed by increasingly cranial horizontal sections, which end at the most cranial convexity of the cerebral cortex. The spin-echo images were obtained at a spacing of 1 cm and a section thickness of 11 mm. The inversion-recovery images cover, in the main, the structures of the brain since, here in particular, additional information is provided. These slices correspond to the sections H9–H15; the thickness is also 11 mm. Figures 6.1 and 6.2 indicate the sectioning directions and the position of all 19 horizontal sections relative to the middle sections of the orthogonal planes, S1 and F1/F9.

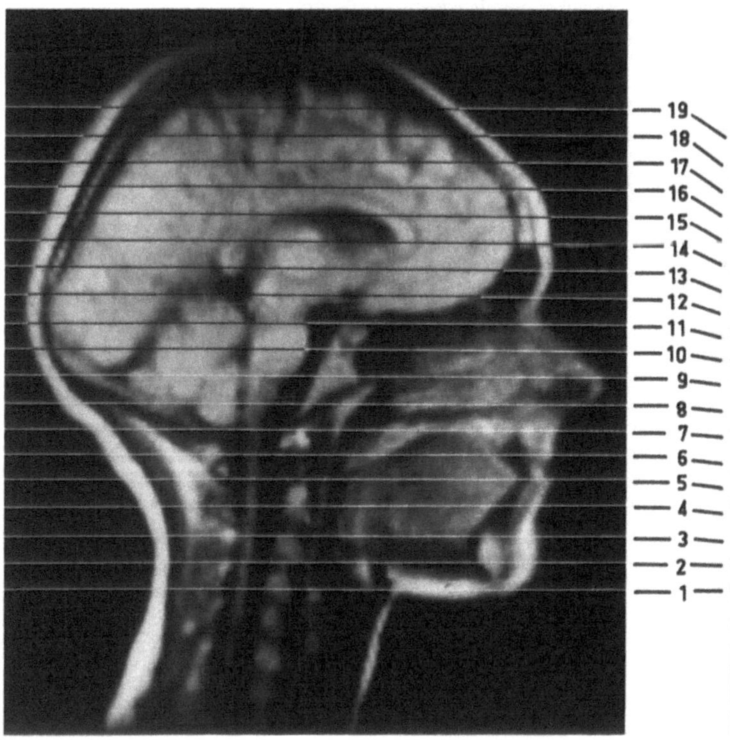

Fig. 6.1

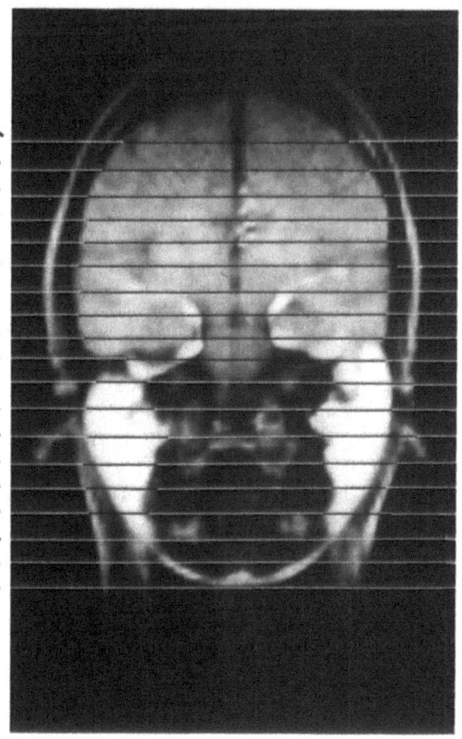

Fig. 6.2

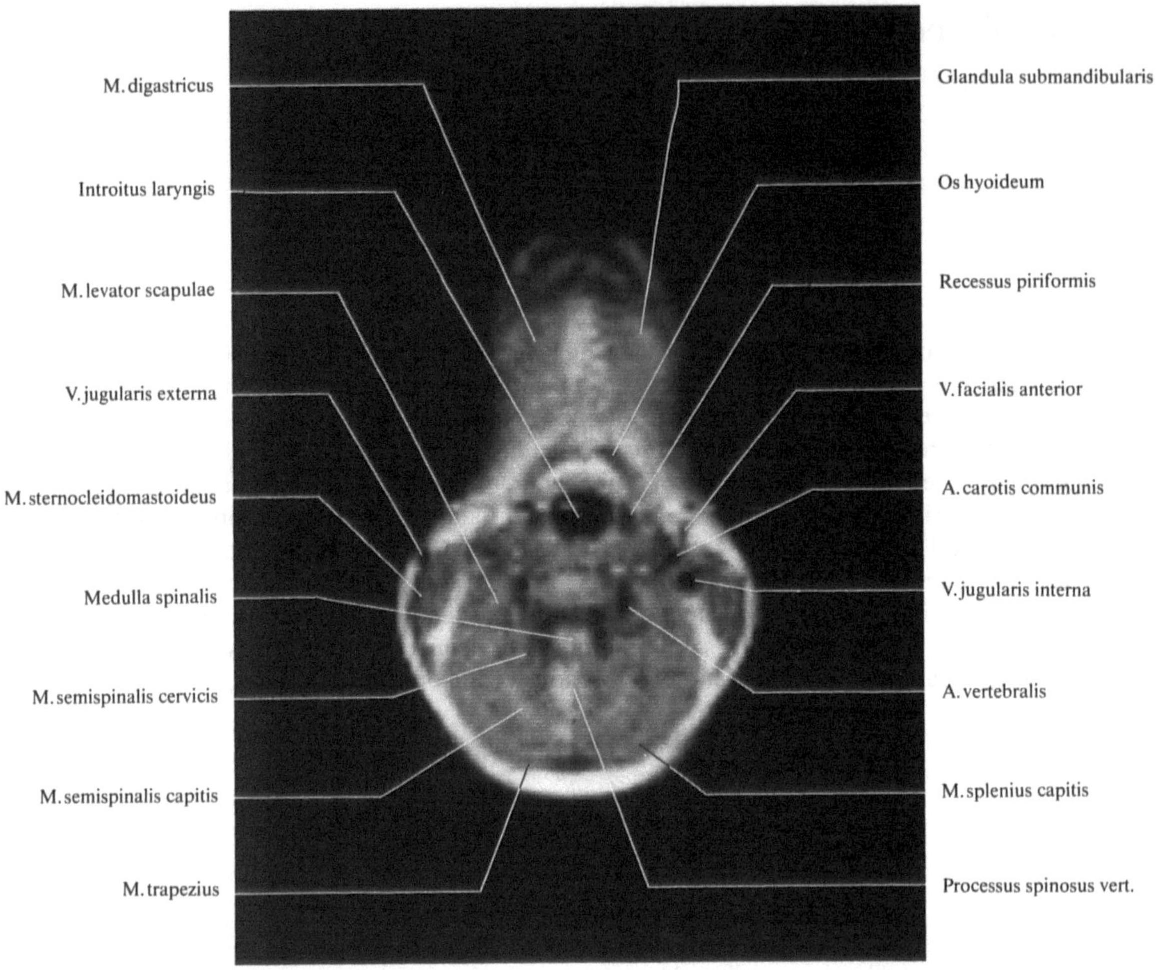

The plane of section H1 is parallel to the chin and passes through the hyoid bone and the upper part of the larynx; these two structures, comprising solid bone and air, are signal-poor and appear black in the image. Medial to the readily definable musculus sternocleidomastoideus the large vena jugularis interna and the narrower carotid artery, located more frontally, can be recognized. These two structures are also visualized black in the image, since the excited protons are carried away by the blood flow. In particular, on the left, the vena jugularis externa, which is subcutaneously in contact with the musculus sternocleidomastoideus, can clearly be distinguished.

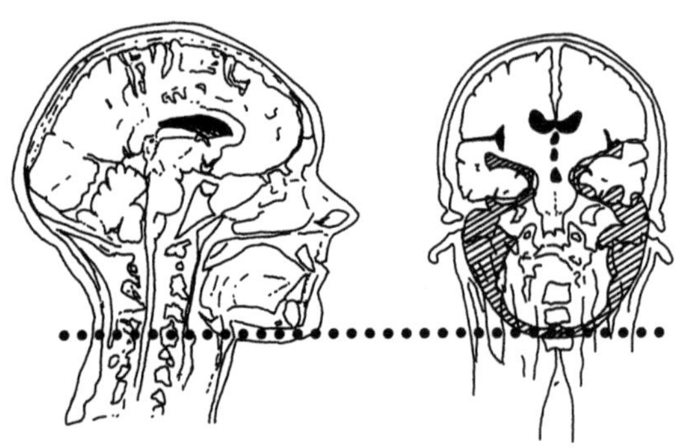

Horizontal section 8 cm below the external acoustic meatus **H 1**

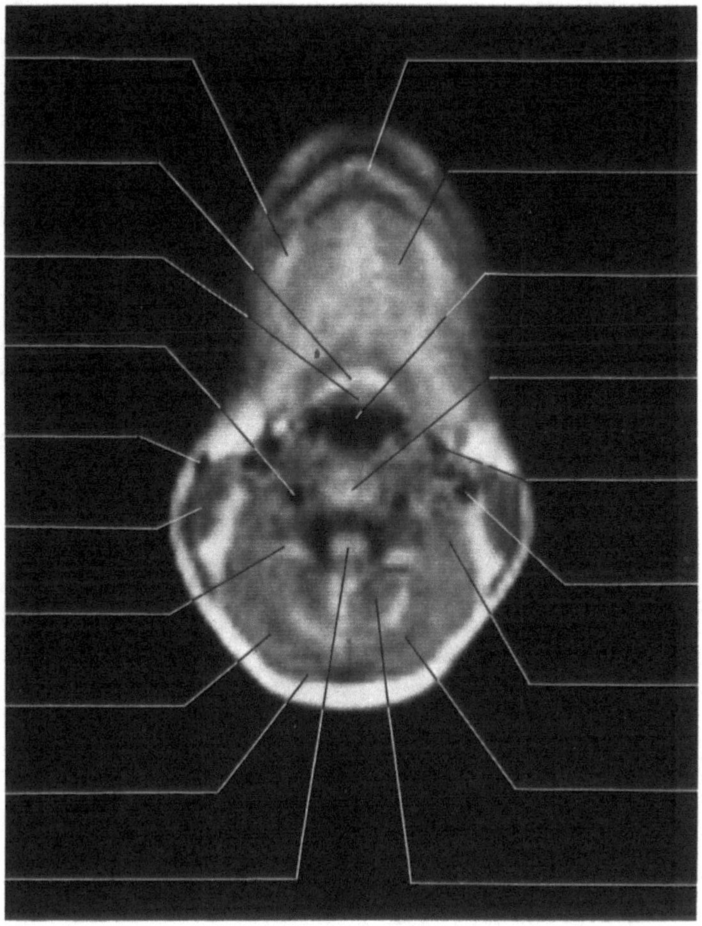

Section H2 passes through the frontal part of the bone-marrow-rich mandible, the epiglottis and the third cervical vertebra. In this section, the great vessels of the neck, as also the arteria vertebralis on both sides, dorsolateral to the vertebral body, can be particularly well distinguished. The most important muscles of the neck, such as the musculi semispinales capitis and cervicis, the musculus levator scapulae, musculus splenius capitis and the narrow musculus trapezius, are all well demarcated by virtue of the fasciae, which appear darker in the image.

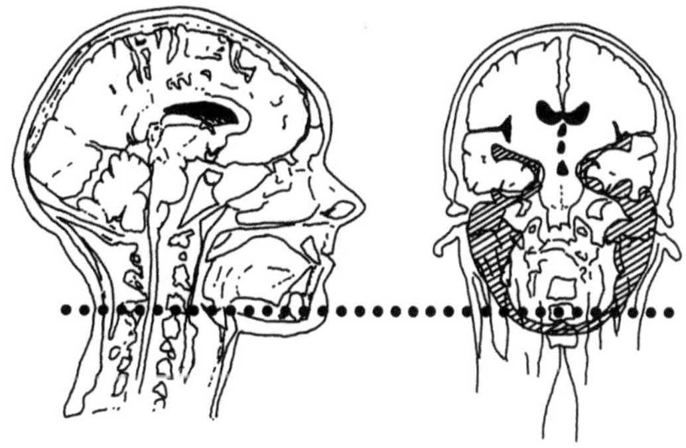

Horizontal section 7 cm below the external acoustic meatus **H 2**

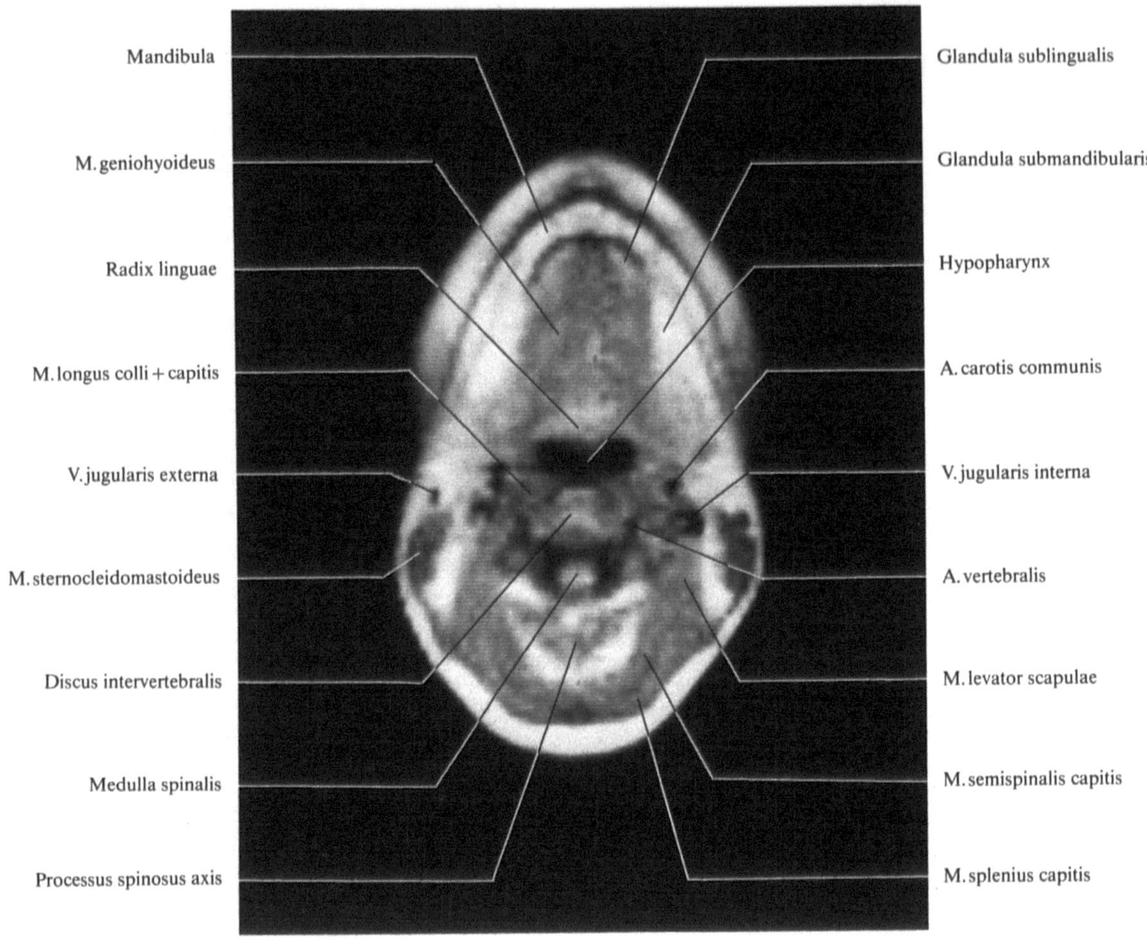

In section H3, dental structures are visible for the first time in the mandible. In the centre of the image the root of the tongue frontal to the hypopharynx, and dorsally, the interventricular disc between the third cervical vertebra and the axis can be seen. The glandula submandibularis and the caudal portion of the glandula sublingualis can be recognized as signal-rich areas. The light-coloured spongy bone of the mandible is framed within the non-signal-generating cortical bone; in front of this the adipose tissue of the face is visualized. The contrasty area lateral to the blood vessels and in front of the musculus sternocleidomastoideus comprises unspecific connective tissue, salivary gland tissue and lymph nodes.

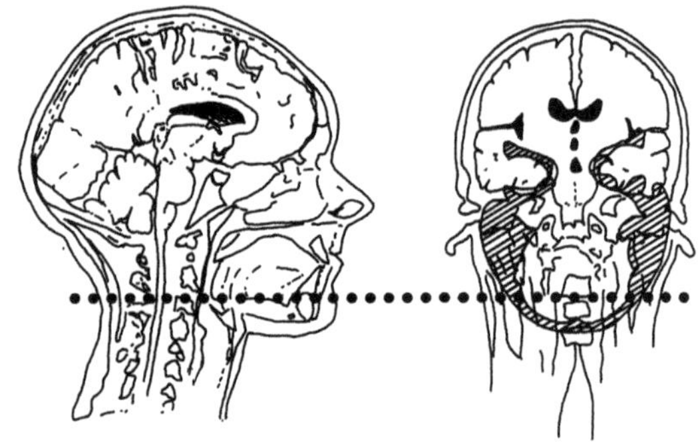

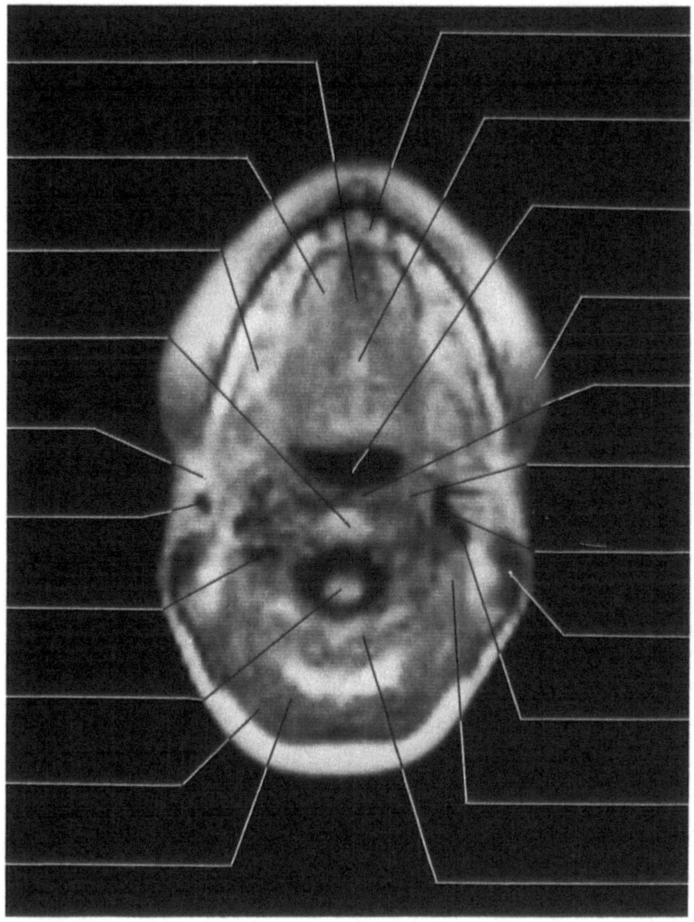

In section H4, the lower row of teeth can be seen, although the teeth are overlapped by the strongly signal-producing tissue of the gums. The lateral portion of the masticatory musculature, the musculus masseter, can be recognized as a darker dorso-lateral area adjacent to the panniculus adiposus. The air-filled space of the pharynx appears as a black area and is separated from the body of the axis vertebra by the muscles of the pharynx, such as the musculus palatopharyngeus and the constrictores pharyngis, and also the anterior muscles of the throat. Laterally, the cord of cervical blood vessels and nerves can be seen, though less readily differentiable than in the preceding sections. The musculus sternocleidomastoideus is visibly tapered and has moved dorsalwards.

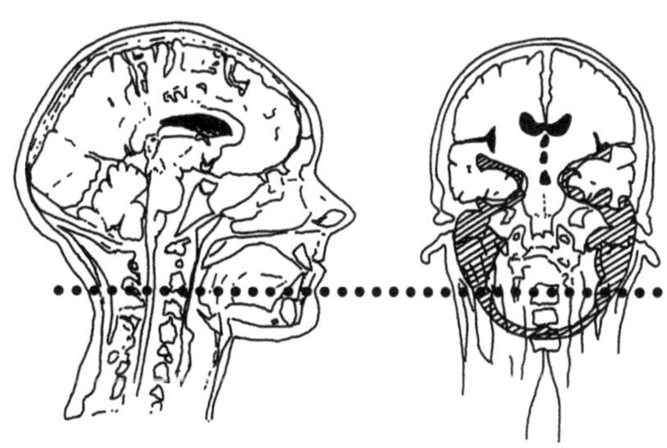

Horizontal section 5 cm below the external acoustic meatus **H 4**

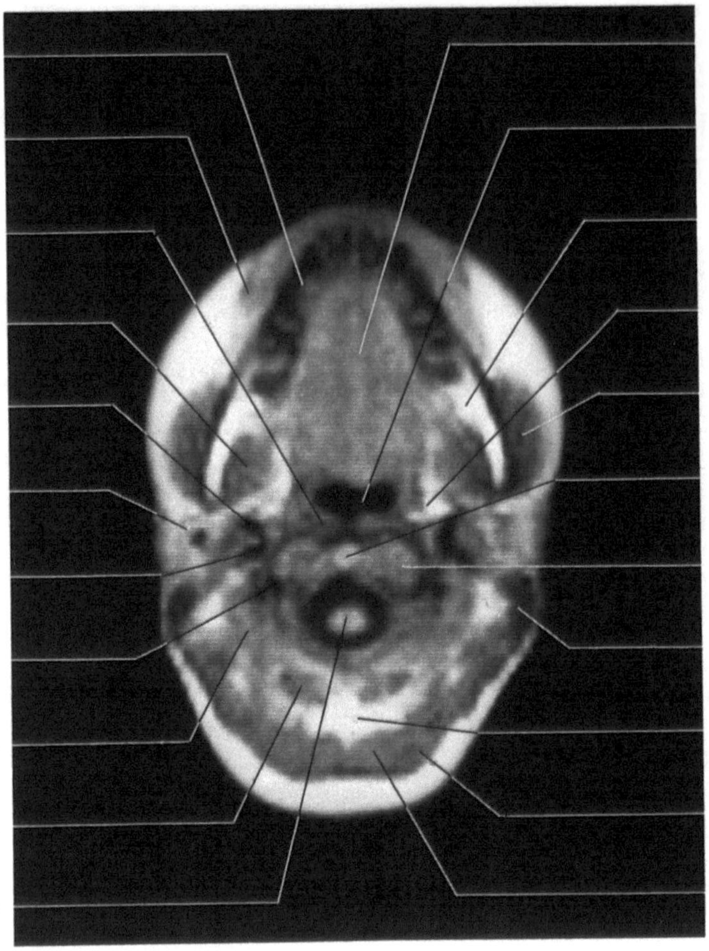

The dental structures visualized in section H5 represent an overlapping of lower and, to a lesser extent, upper teeth. Their light extension towards the dorso-lateral is the angle of the mandible; the lateral, dark layer of cortical bone is well marked. The musculus masseter and small facial muscles, located frontally, are readily distinguishable from the surrounding fatty tissue. Medial to the angle of the mandible medial aspects of the masticatory musculature, the musculus pterygoideus medius, can be seen. The round dens of the axis contrasts well with the bony structures formed from the atlas and axis.

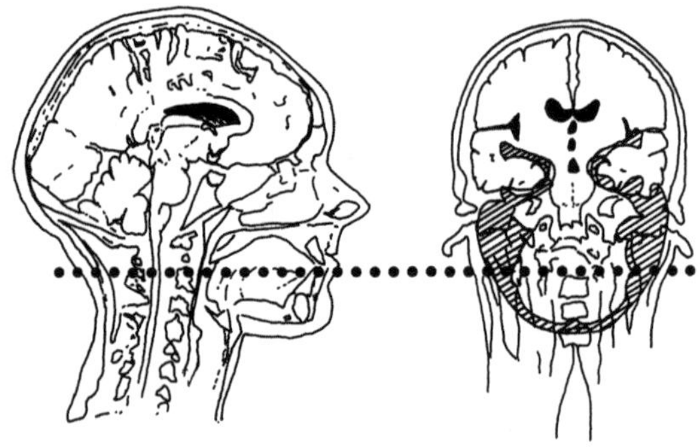

Horizontal section 4 cm below the external acoustic meatus **H 5**

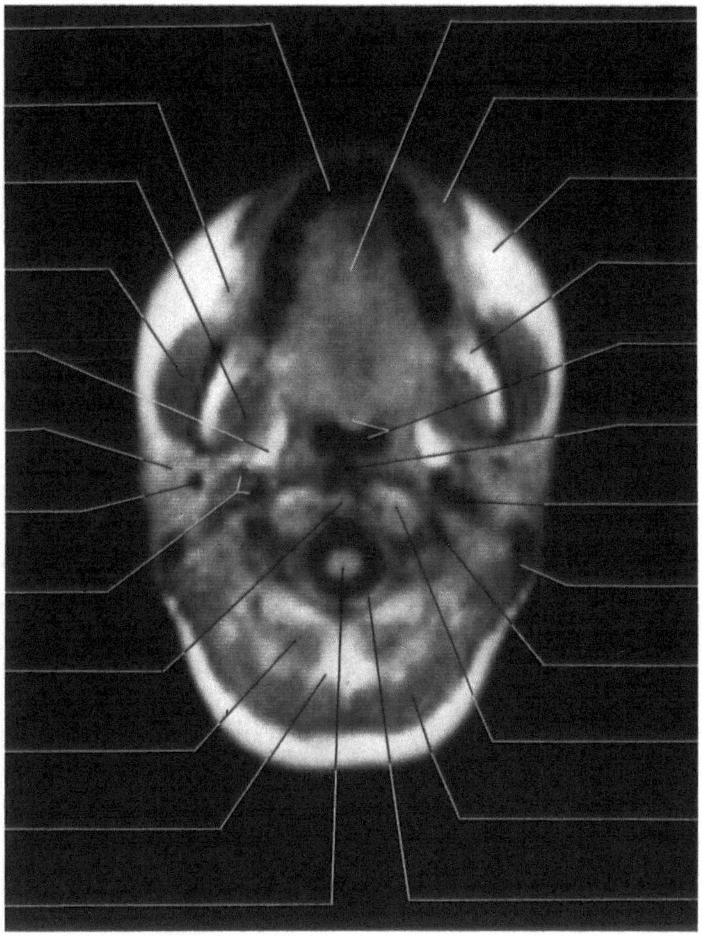

Section H6 passes exactly through the middle of the upper dentition, which, owing to their oblique position, produce a larger projection on to the plane of the image and thus appear more massive than the lower teeth. As part of the soft palate the uvula projects into the pharynx. The salivary gland, in the middle of which the vena jugularis externa is located, is well demarcated from the surrounding tissue. The intensity of the signal is approximately between that of muscle and fat. The dark cavum subarachnoidale, which can be differentiated around the spinal cord, expands to form the foramen magnum, seen in the following section. A predominant feature in the neck is the light ligamentum nuchae, which encloses the posterior straight muscles (musculus rectus capitis posterior) of the neck. The musculus trapezius, with its tendinous insertion, can no longer be differentiated from the scalp.

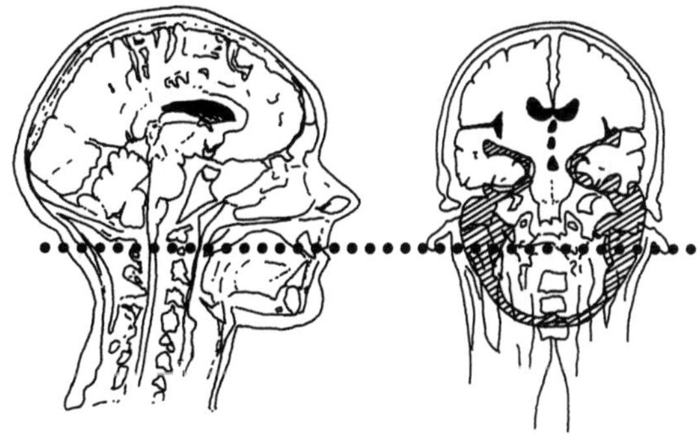

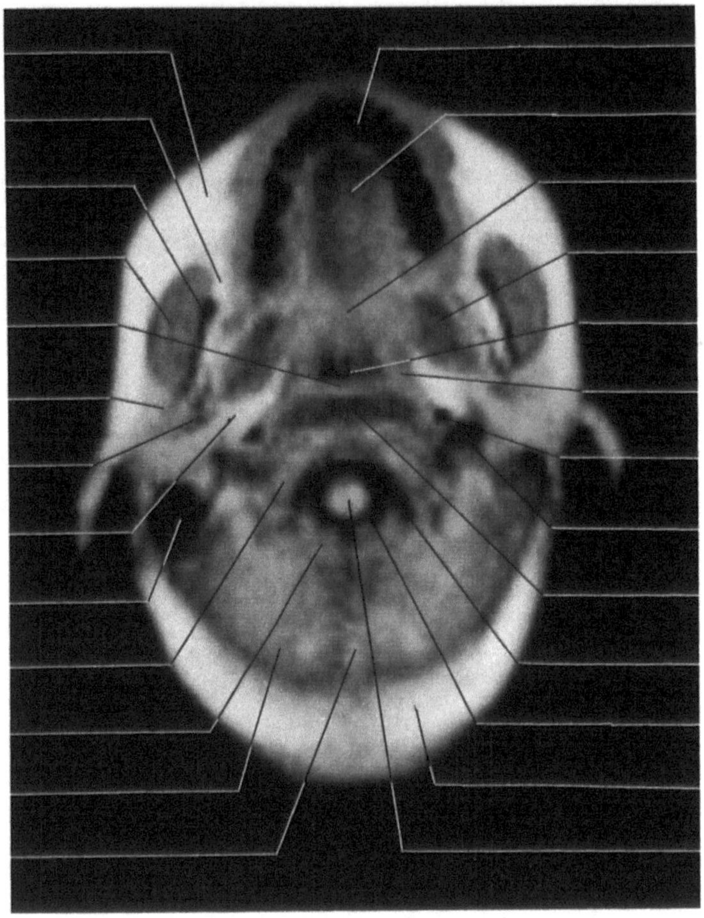

In section H7 the frontal portion of the image reveals the roots of the upper teeth, together with the hard palate; dorsally the foramen magnum may be seen, the posterior boundary of which, however, is no longer visualized, being "replaced" by the tonsil of the cerebellum. The light area dorsal to this represents an overlapping of cisternae, occipital bone, nuchal fat and muscle. The ligamentum nuchae is distinguishable as a trident-shaped structure, the prongs of which are directed dorsally. Particularly easy to recognize are the two carotid arteries lateral to the musculi longi colli and the basal portion of the os occipitale.

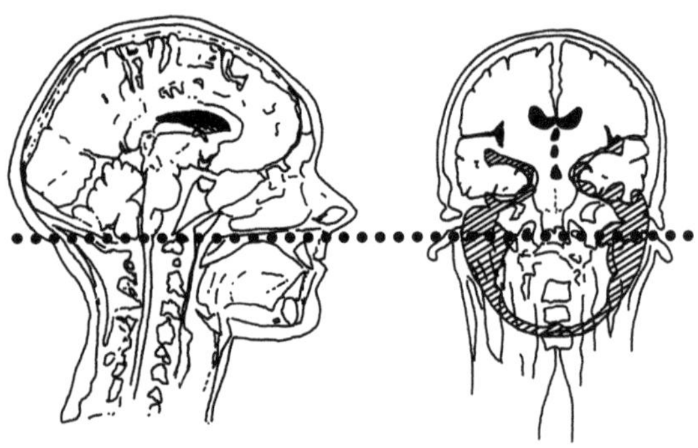

Horizontal section 2 cm below the external acoustic meatus **H 7**

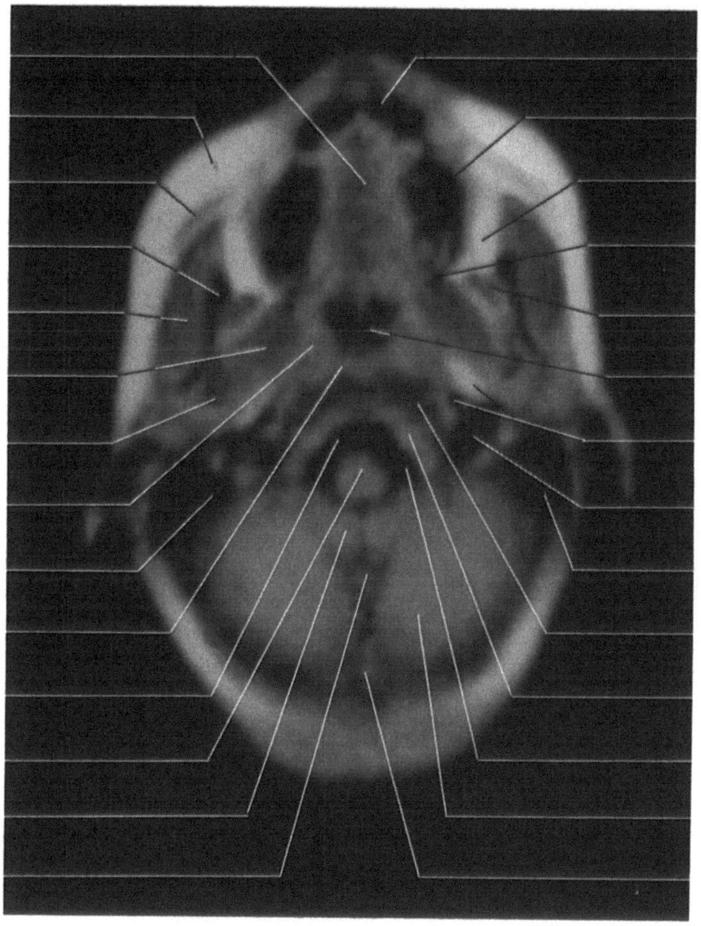

Section H8 runs from the processus alveolaris of the upper jaw (maxilla) along the palate, through the tonsillar pharyngea, the basilar portion of the clivus and the medulla oblongata to the cerebellum. In the maxillary sinus structures of the dental alveolae can be seen, while at their dorsal ends the processus pterygoidei can be recognized. The slender ribbon of the cortical bone of the mandible is framed by the two large masticatory muscles; fronto-medially, the insertion of the musculus temporalis can be seen. The arteria carotis interna and the vena jugularis interna, which is sectioned through its bulb, are readily recognizable, particularly on the right. The two vertebral arteries have not yet joined to form the arteria basilaris; they appear, surprisingly well differentiated, in the cisterna basilaris.

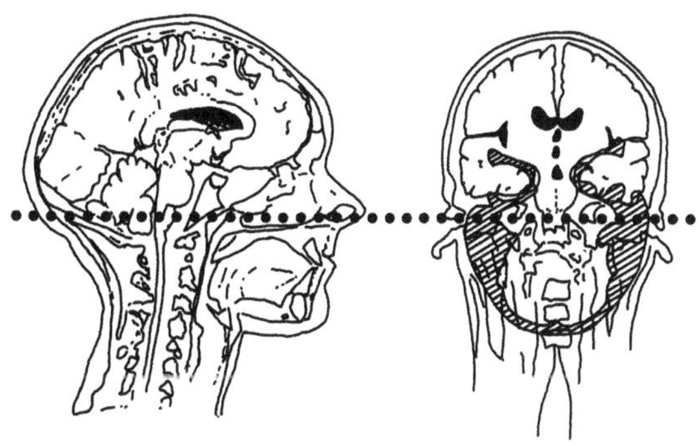

Horizontal section 1 cm below the external acoustic meatus **H 8**

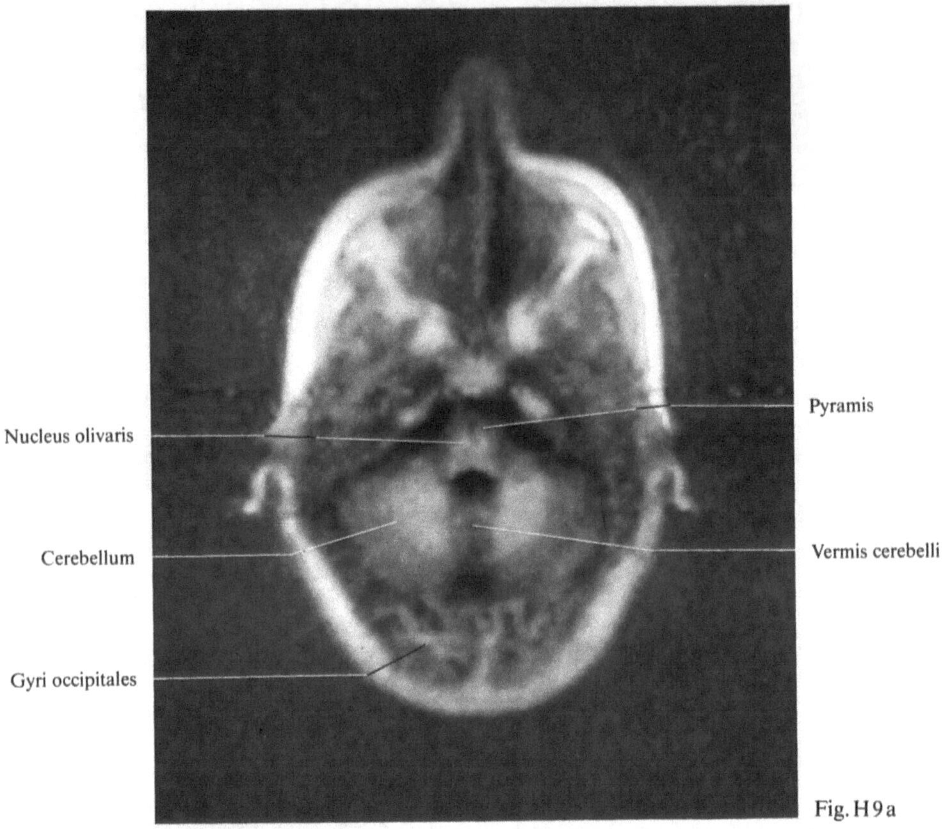

Fig. H9a

Section H9 passes approximately through the external auditory meatus and, dorsally, just cuts the caudal portion of the occipital lobe. The predominant feature, however, is the cerebellum located in the posterior cranial fossa (fossa posterior). In the inversion-recovery image H9a the differences between the more layered, concentric structure of the cerebellum and the convolutions and sulci of the cortex, as exemplified in the occipital lobe, can be recognized particularly well. In the medulla oblongata too Fig. H9a reveals finer details such as the pyramidal tracts and the olive.

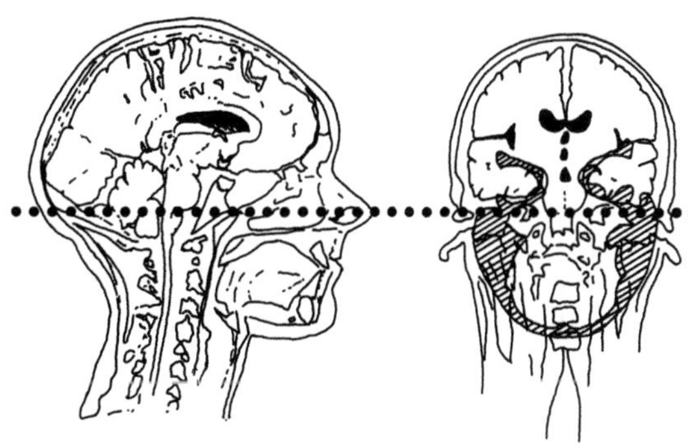

Horizontal section through the external acoustic meatus **H 9**

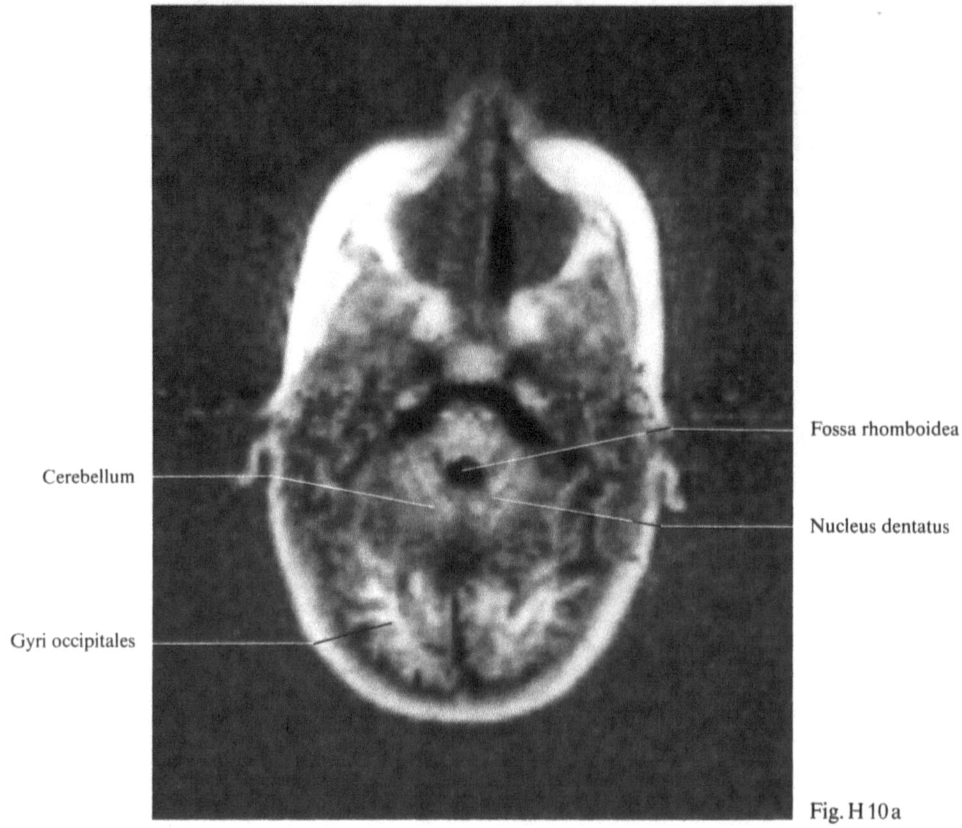

Fig. H 10 a

Section H10 passes from the middle part of the bridge of the nose at the level with the floor of the sella through the pons. In the sinus maxillaris, on the left, the ductus nasolacrimalis is clearly visualized. In the inversion-recovery image H10a, in contrast, the cerebellum and the occipital lobes are well differentiated on the basis of their different appearance. While in the spin-echo image the muscles, fasciae and bony structures in the region ventral to the posterior fossa are clearly differentiated. H10a reveals only few image-generating contrasts. In contrast to the spin-echo image, here the internal and external CSF spaces, i.e. the fourth ventricle and the basilar cisternae, appear much more distinct than air and bone.

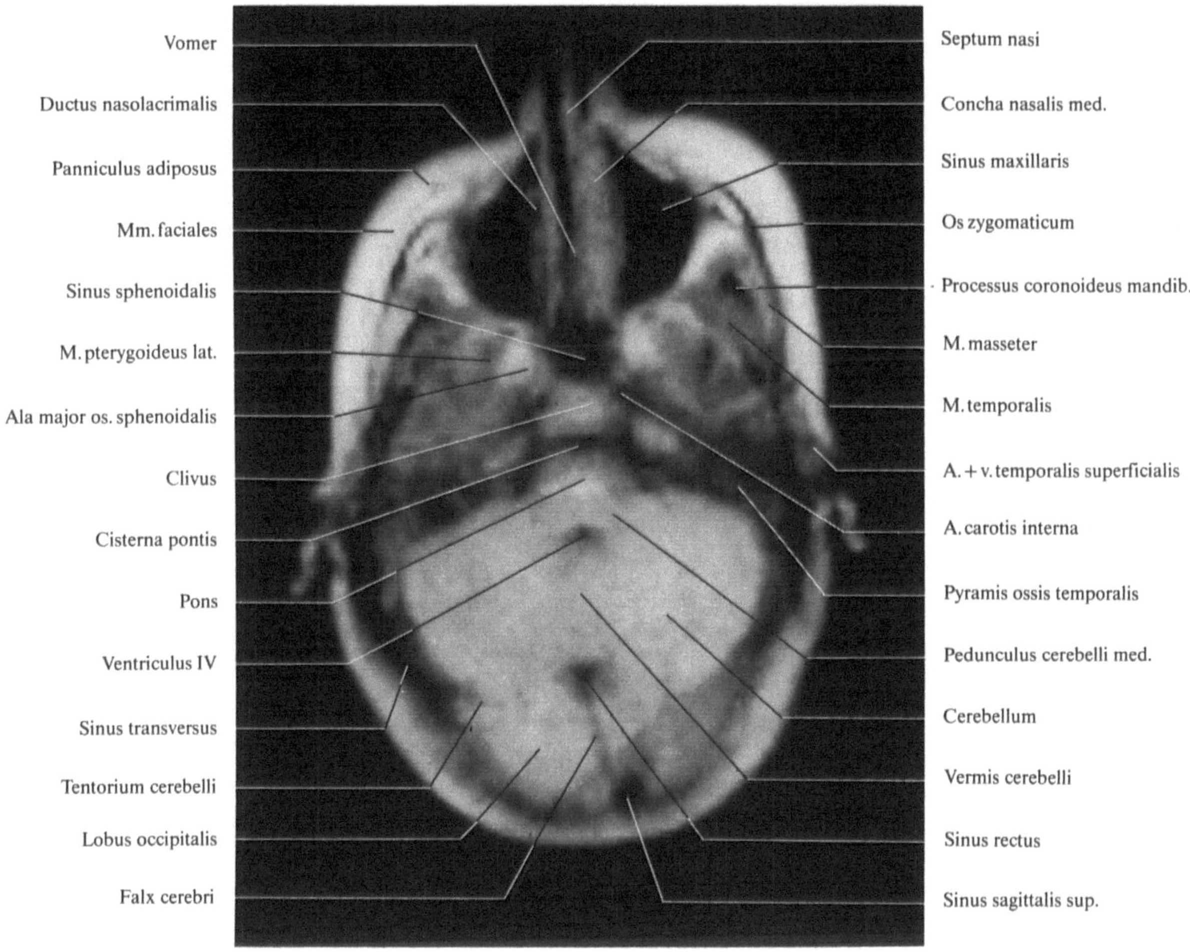

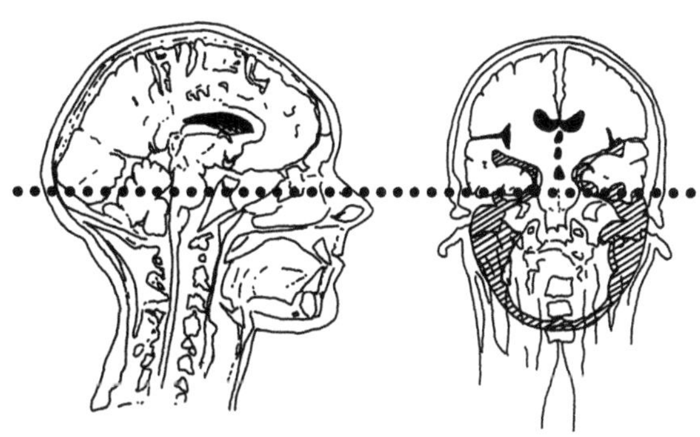

Horizontal section 1 cm above the external acoustic meatus **H 10**

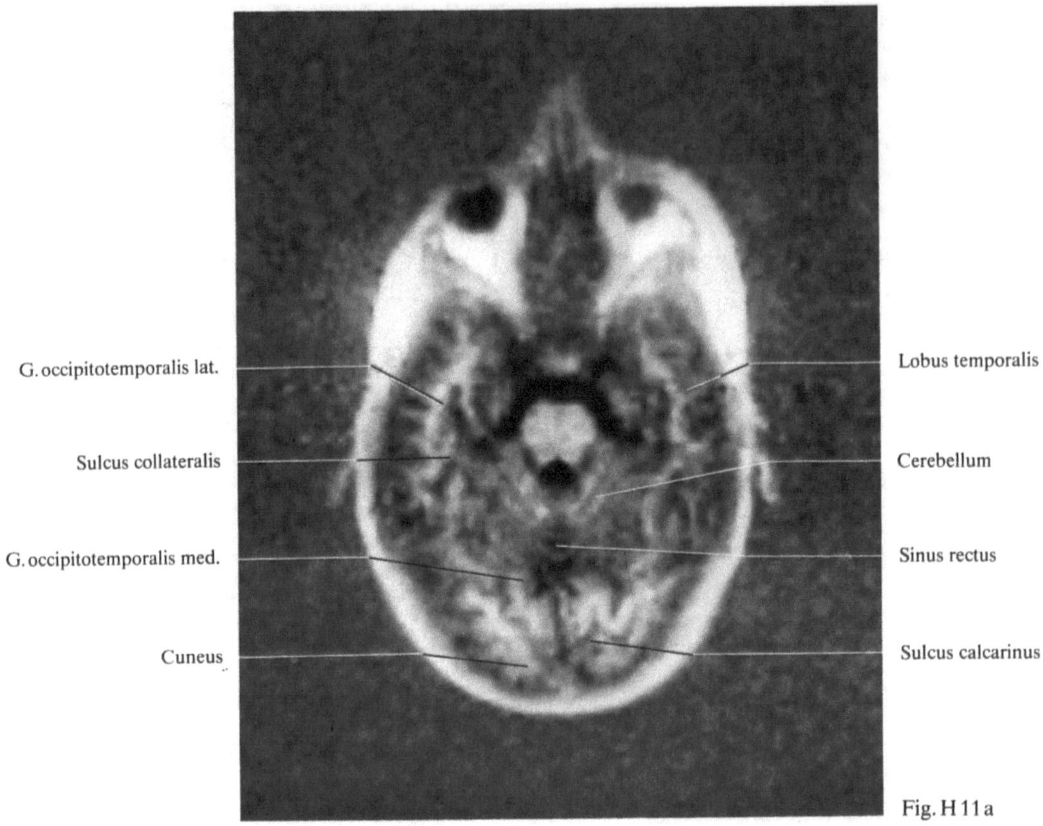

Fig. H 11 a

Section H11 cuts the floor of the middle cranial fossa (fossa cranii media), where the temporal lobe is just visualized. Ventrally, orbital structures can be recognized in the spin-echo image, particularly the orbital fat. The corresponding inversion-recovery image H11 a is only slightly more cranial and provides a more contrasty visualization of the vitreous body of the eye. The convolutions and sulci of the frontal and occipital lobes appear with greater clarity. The tissue of the cerebellum, located between the aqueduct and the sinus rectus, can be distinguished from that of the cortex only in the inversion-recovery image H11 a.

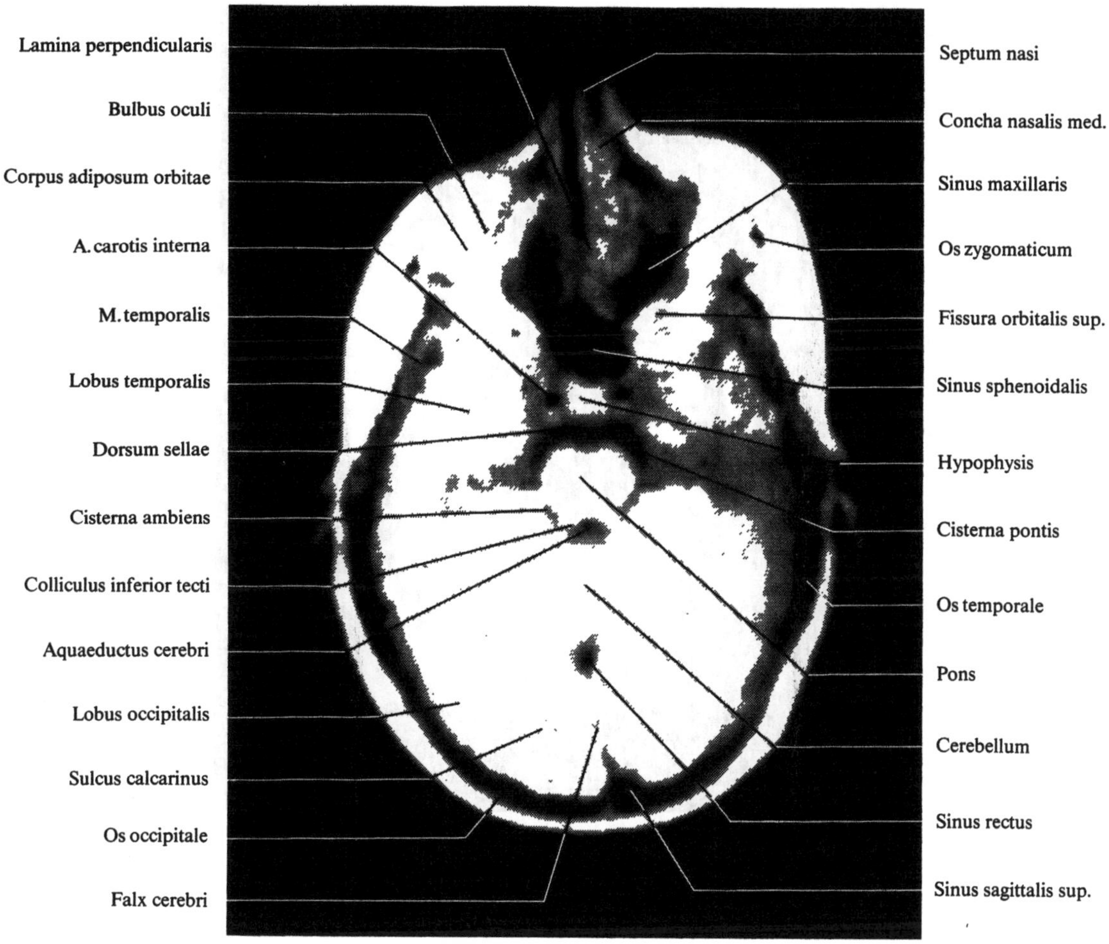

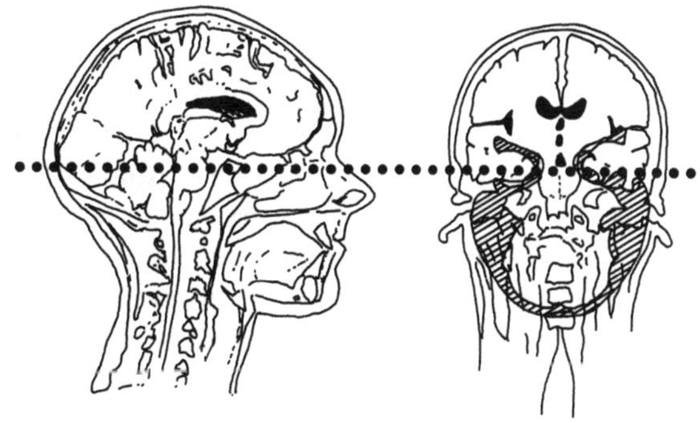

Horizontal section 2 cm above the external acoustic meatus **H 11**

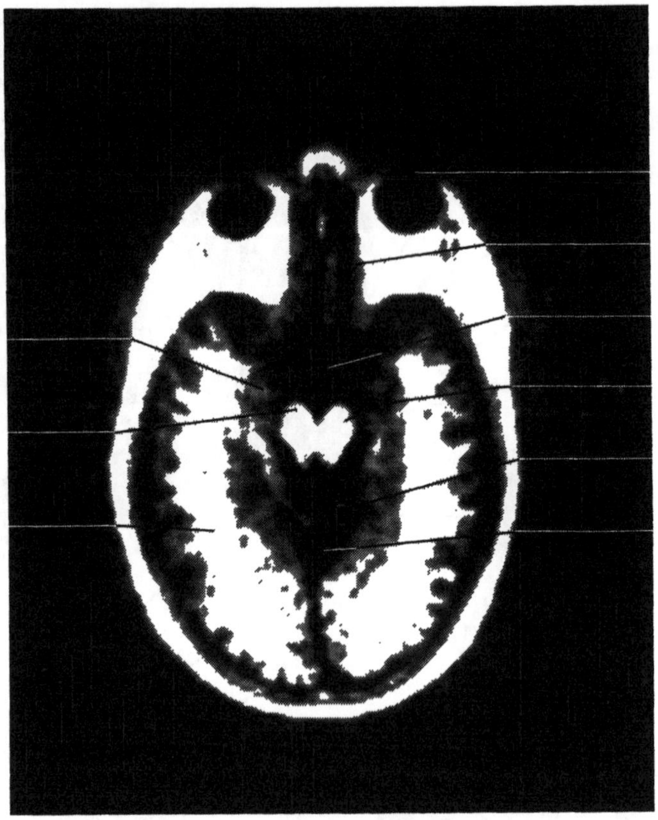

Fig. H12a

Section H12 passes almost exactly through the centre of the eyeballs. The anterior chamber and the lens are considerably better visualized in Fig. H12a, the inversion-recovery image, while in the spin-echo image the optic nerve, together with the optic chiasma and the rectus muscles (musculi recti) of the eye, can be more readily discerned. Dorsal to the optic chiasma the narrow sagittaly directed recessus infundibuli can be recognized; further dorsally appears the midbrain, with such formations as the crus cerebri, the tectal (quadrigeminal) lamina and the nucleus ruber, which in Fig. H12a is visualized somewhat darker than its surroundings. Laterally, the uncus of the gyrus parahippocampalis can be readily recognized. In the T_1 enhanced image, the dorsally running optic radiation can be distinguished. Additional images were obtained with the same direction of section to provide pictorial visualization of the relaxation times T_1 and T_2, (see Fig. 2.4a–c). They are discussed in detail in Chap. 2.

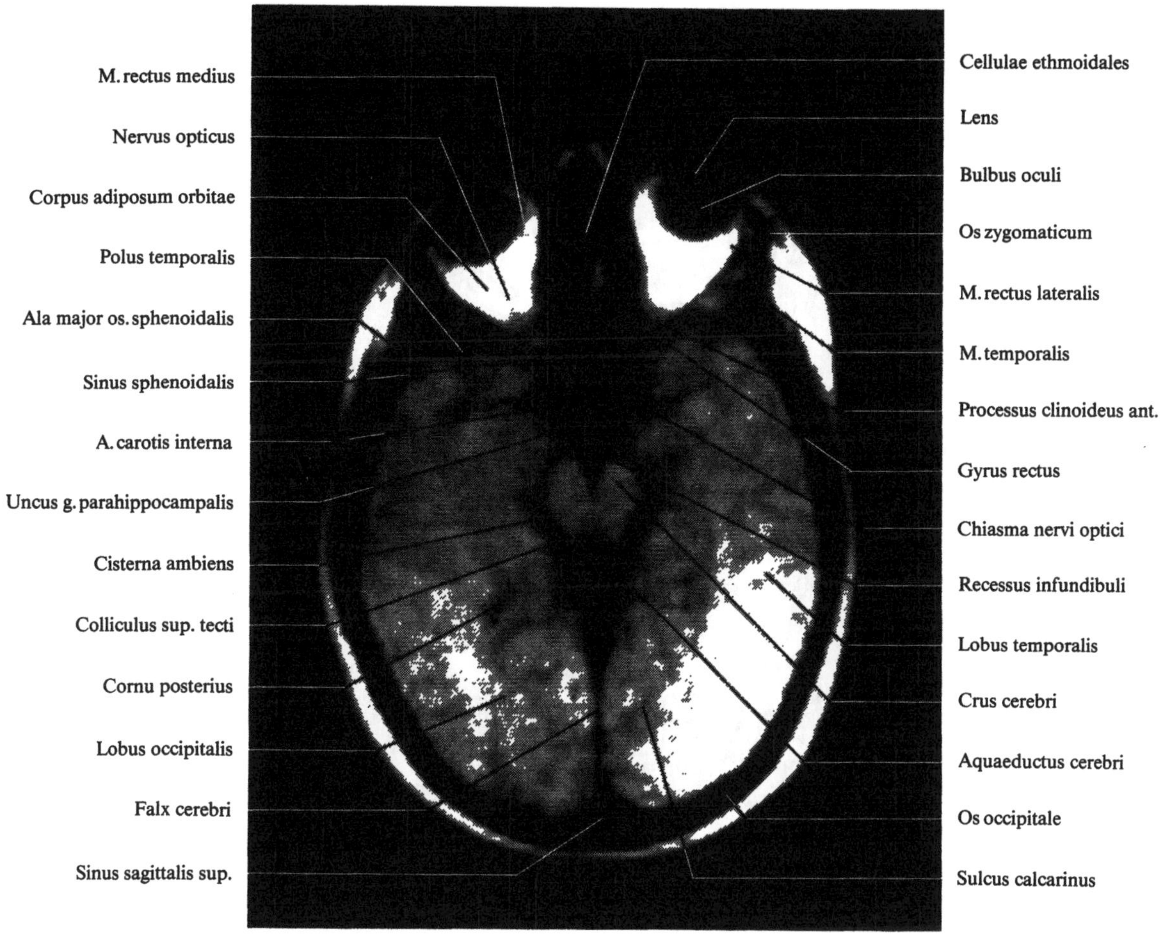

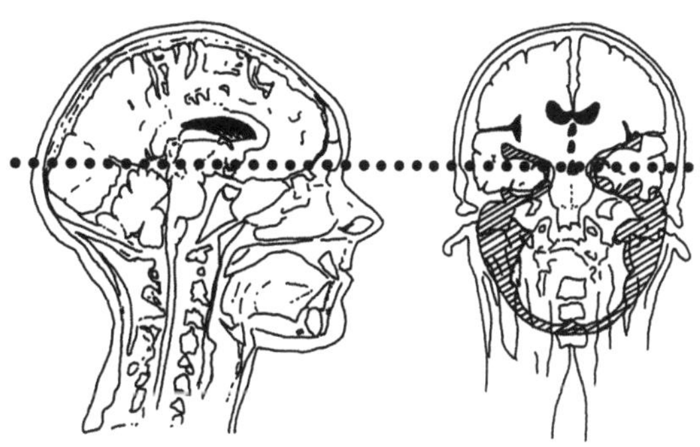

Horizontal section 3 cm above the external acoustic meatus

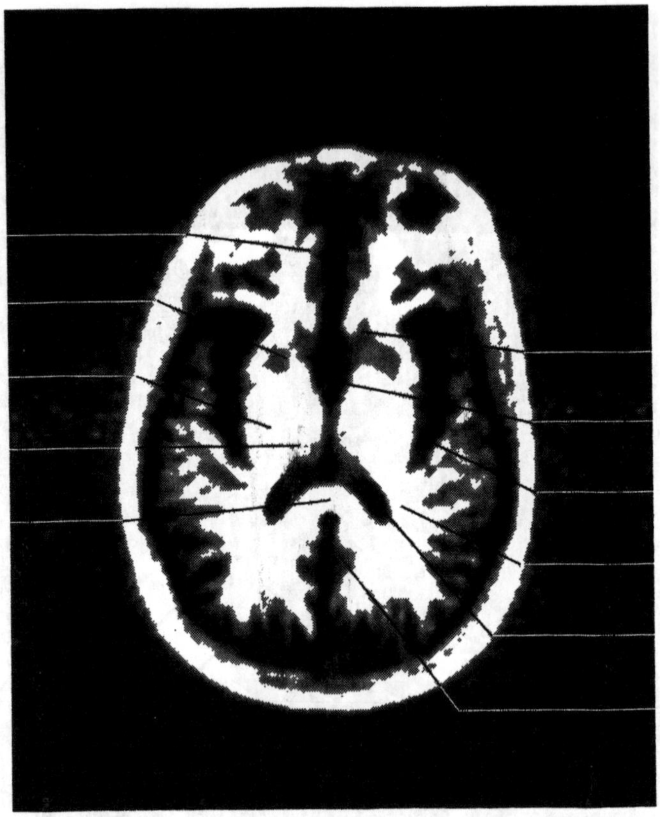

Fig. H 13 a

In section H13 (and in particular in the inversion-recovery image H13 a) the cranial portion of the orbit, the initial structures of the frontal part of the brain (gyrus rectus), the temporal lobe with the sulcus lateralis, the nuclei of the diencephalon and the ventricles are all particularly well visualized. The nucleus lentiformis is very clearly demarcated from the white matter of the brain, for example, the internal capsule. The air-filled frontal sinus is much more sharply visualized in the spin-echo image, while the CSF spaces are more readily recognized in the image H13 a.

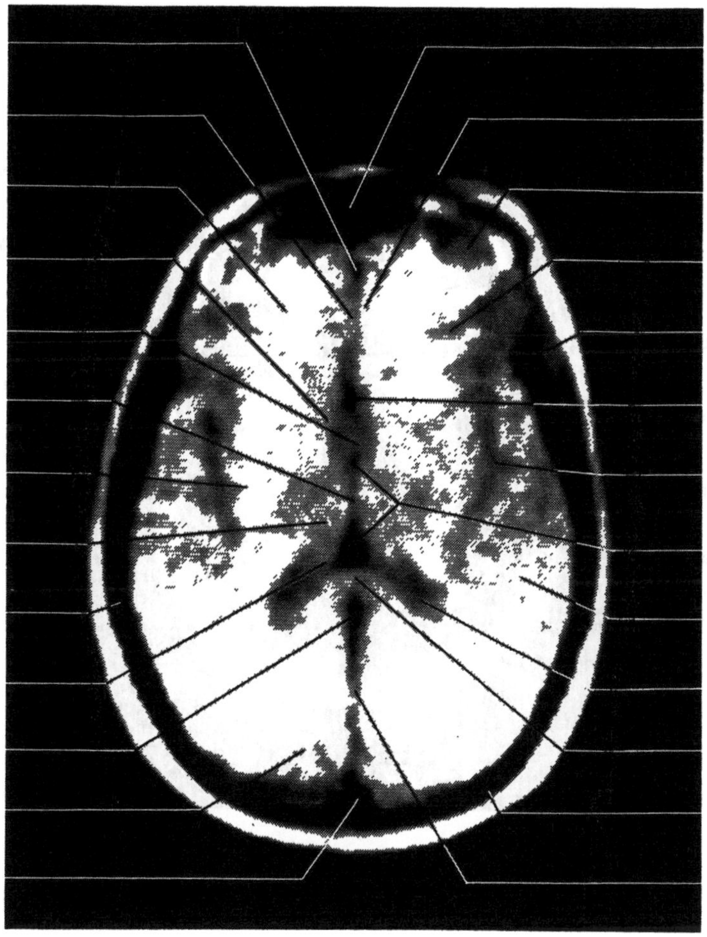

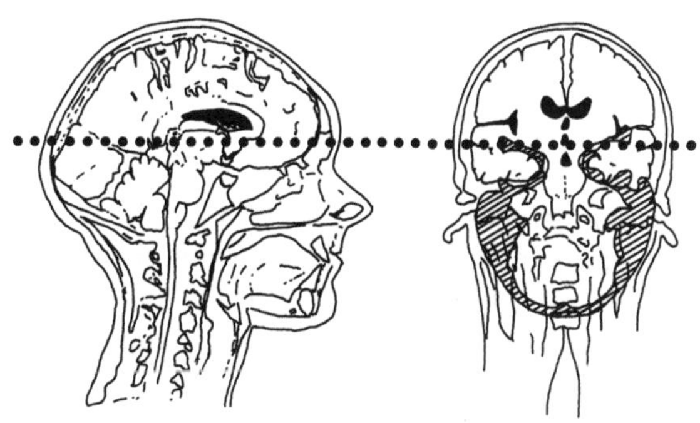

Horizontal section 4 cm above the external acoustic meatus **H 13**

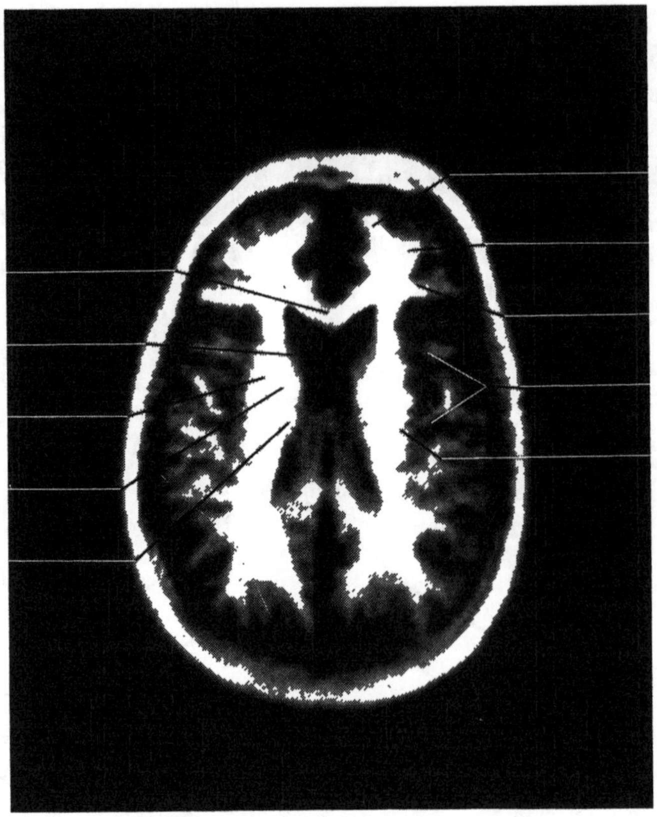

Fig. H 14 a

The main structures in section H14 are the two lateral ventricles together with their anterior horns, the pars centralis separated by the septum pellucidum and the atrium. In the inversion-recovery image H14a the close proximity of the caput nuclei caudati to the anterior horn can be recognized; the former is visualized darker than comparable structures, such as the thalamus and the lentiform nucleus, because of overlapping with the CSF. The readily discernible internal capsule separates these three structures from one another, while the external capsule demarcates the undifferentiable claustrum from the insula. A discussion of the images obtained with the identical section position but variations in the technical parameters *T* and *d* is found in Chap. 2 Fig. 2.2 a–f.

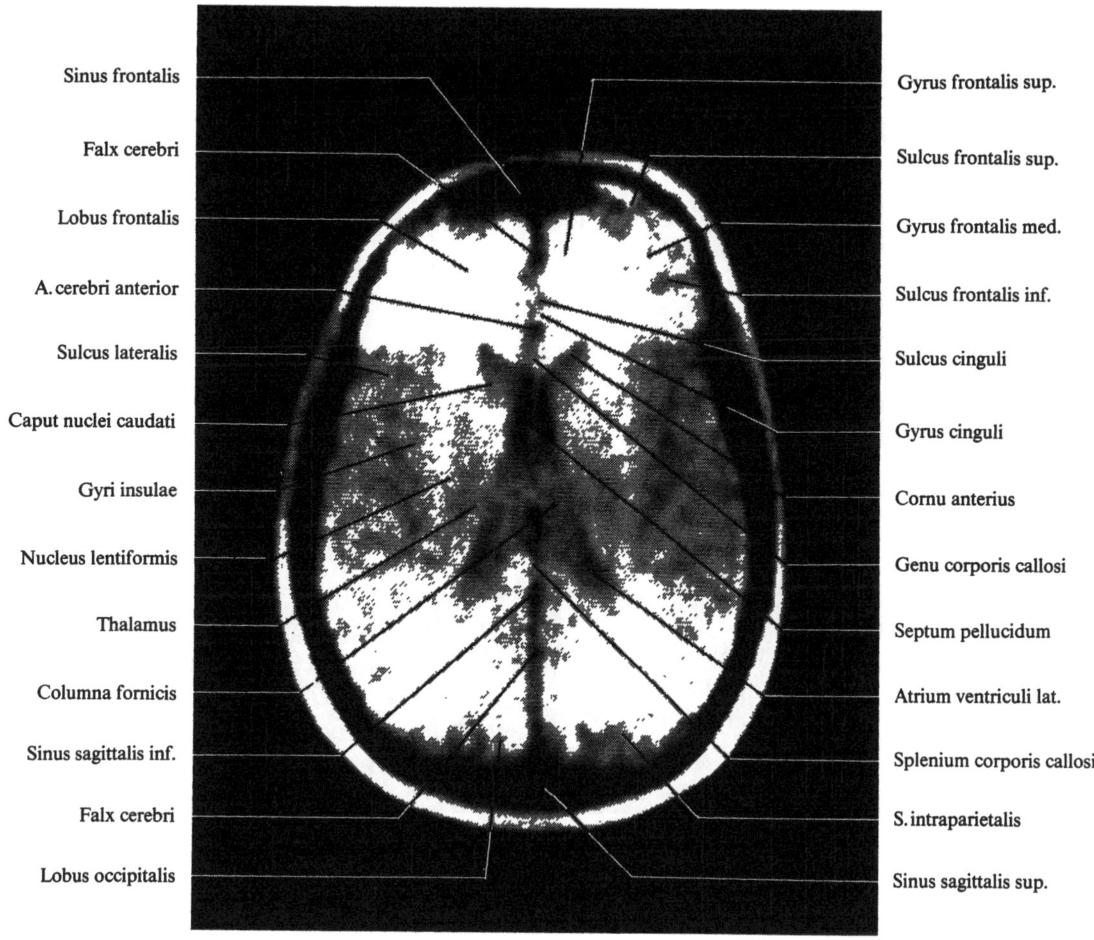

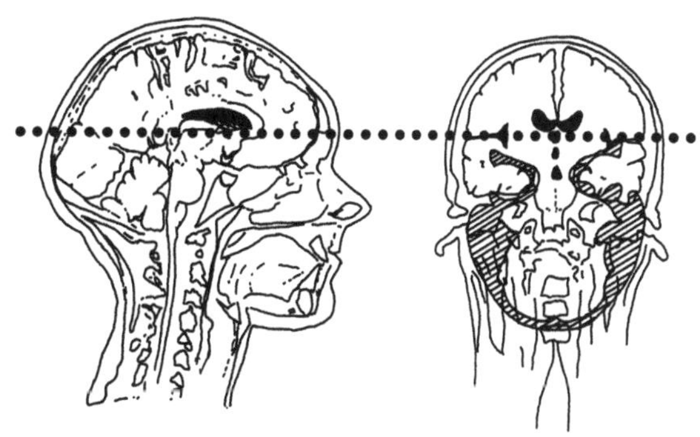

Horizontal section 5 cm above the external acoustic meatus **H 14**

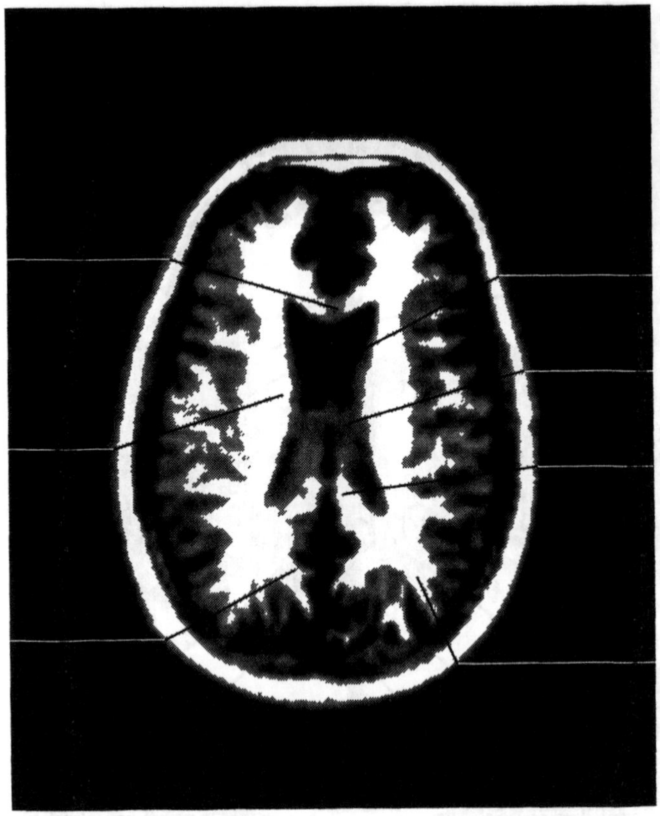

Fig. H 15 a

The spin-echo image and the inversion-recovery image (H15 a) of the section H15 are probably shifted relative to each other by about 5 mm in the vertical plane, the latter being located more cranially. This can be particularly well recognized in the case of the lateral ventricles, which in H15 in H15 a the atria can also be recognized. In the diencephalon (interbrain) the nucleus caudatus and the thalamus can be distinguished. The convolutions and sulci of the cerebral cortex are now more clearly distinguishable.

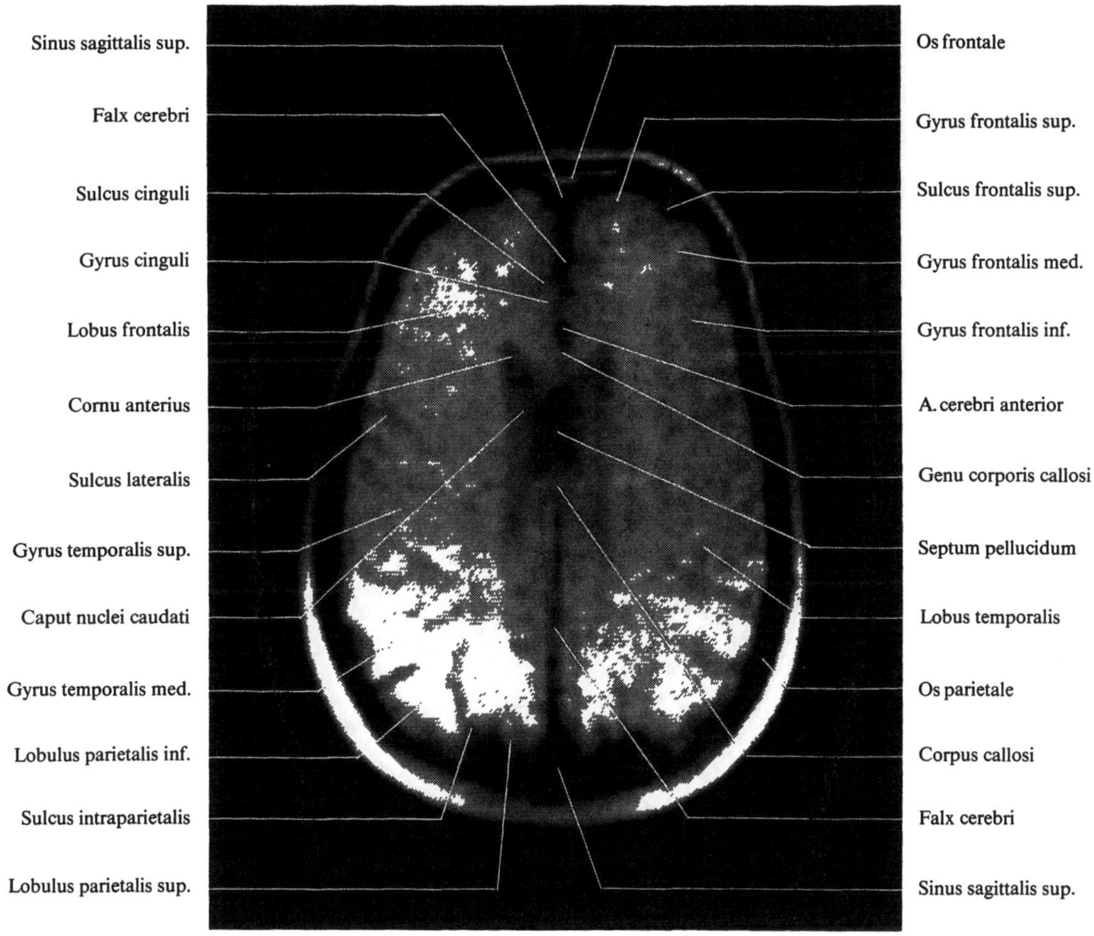

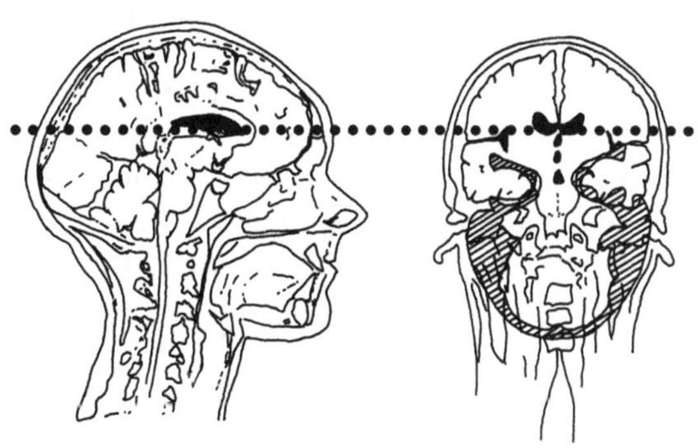

Horizontal section 6 cm above the external acoustic meatus **H 15**

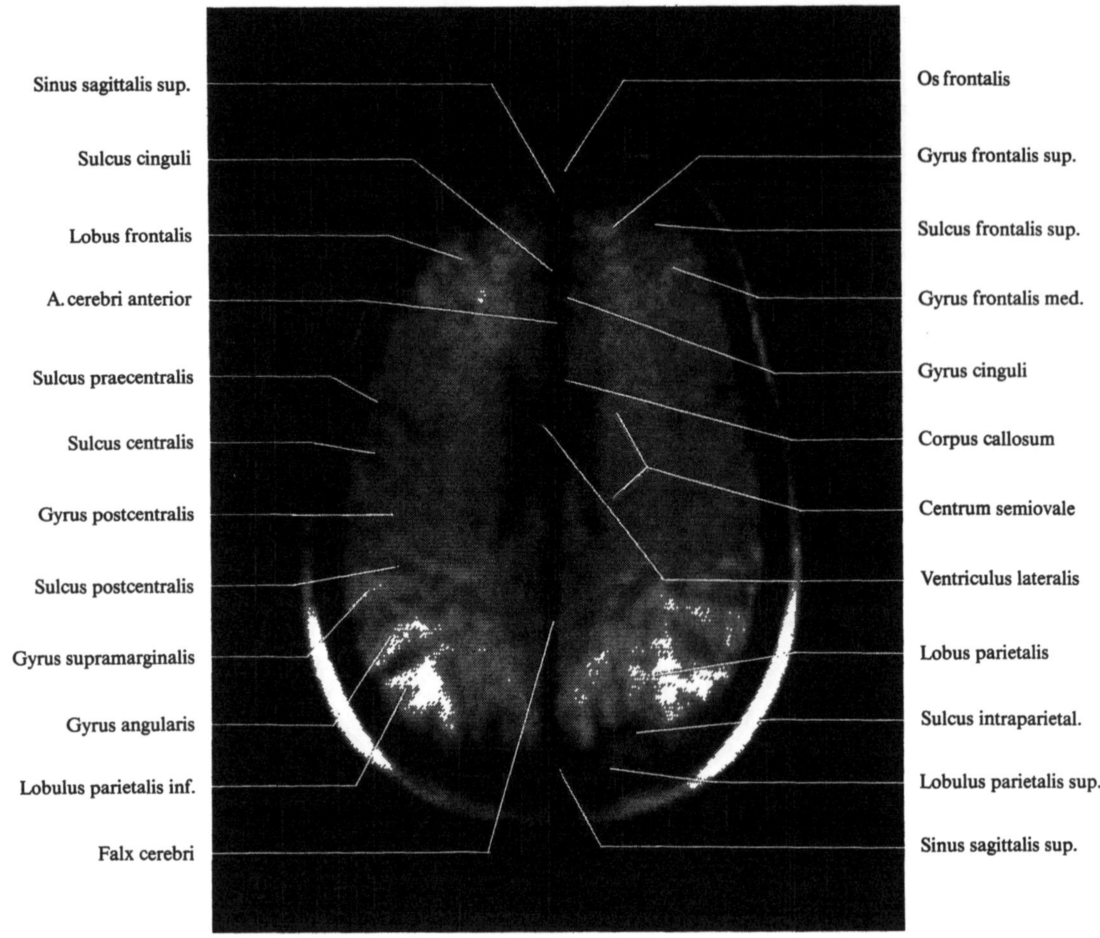

Section H16 is the highest horizontal slice in which the ventricles can still be seen. The corpus callosum is overlapped by the sinus sagittalis inferior and the falx cerebri, so that the two cerebral hemispheres appear completely separated. Where in section H15 the sulcus lateralis was to be seen the sulcus precentralis now appears together with the dorsally located adjacent lobulus paracentralis. In the lower third of the image the gyrus angularis and the gyrus supramarginalis, which represent the frontocaudal transition from the parietal to the temporal lobe, can be made out.

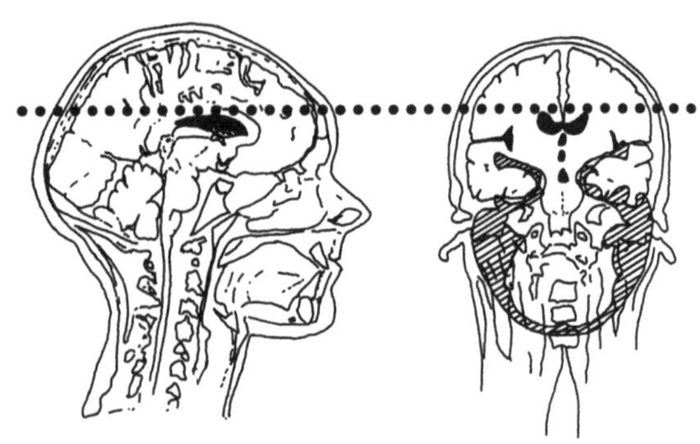

Horizontal section 7 cm above the external acoustic meatus **H 16**

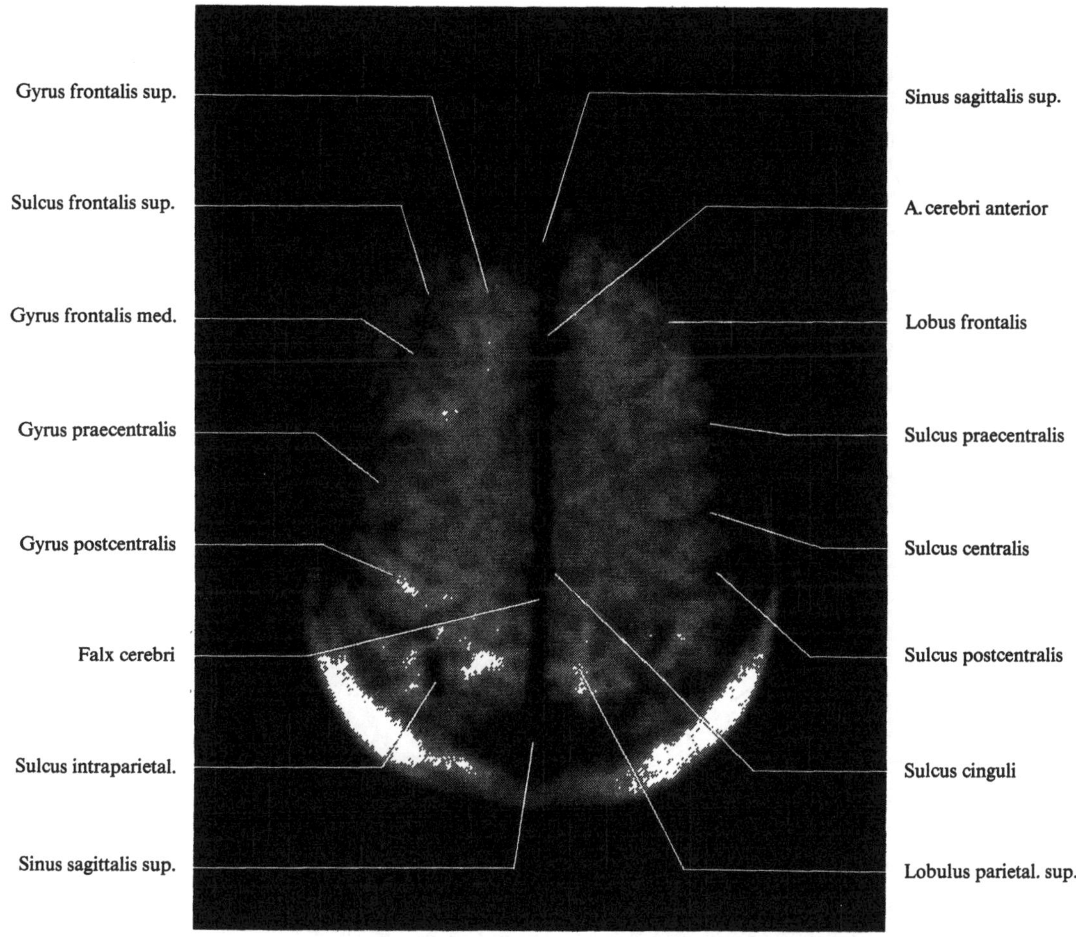

Section H17 shows improved contrast between the convolutions and the sulci. The various directions of the structures of the cerebral cortex are now more clearly determinable. Thus, in the frontal and parietal lobes the gyri follow a more sagittal direction, while those in the intervening lobulus paracentralis take a largely coronal course. As a result the individual areas are fairly readily defined. The sinus sagittalis superior appears, both frontally and dorsally, as a dark extension of the falx cerebri. The epicranial aponeurosis (galea aponeurotica), which is richer in signals, increasingly covers the bones of the skull owing to their oblique projection.

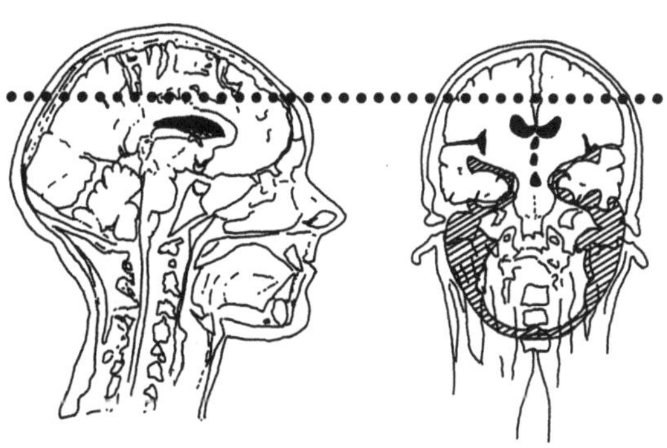

Horizontal section 8 cm above the external acoustic meatus **H 17**

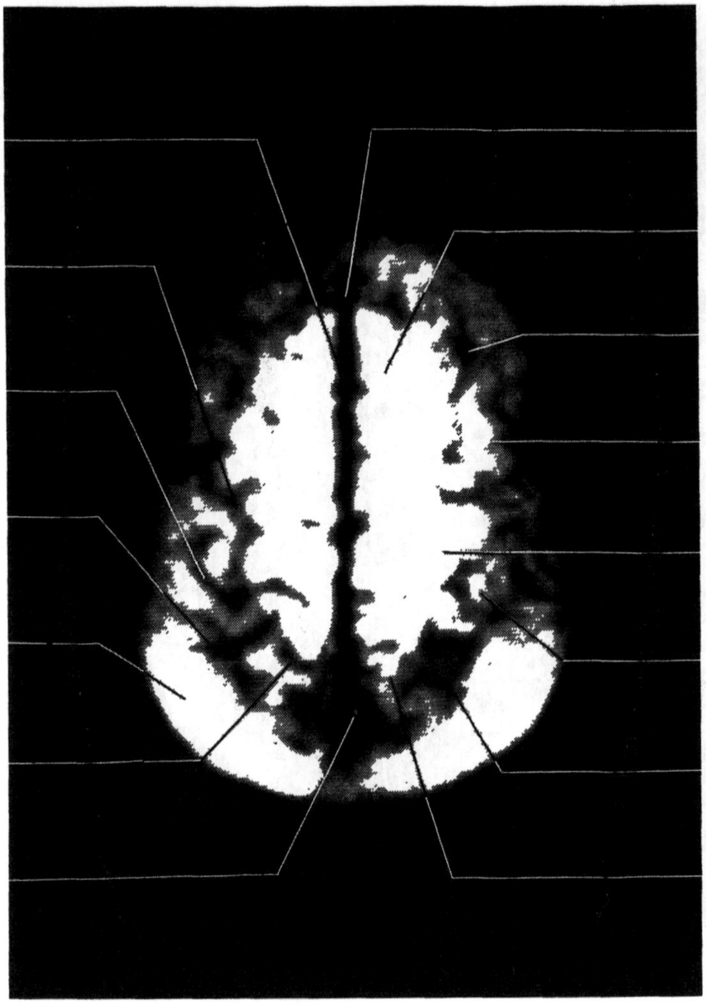

In the section H18 the frontal lobe is composed almost entirely of the gyrus frontalis superior, while the gyrus frontalis medius disappears among the brain tissue, parietal bone and epicranial aponeurosis. The sulcus here designated as sulcus cinguli has various names in different nomenclatures; it is a cranially directed prolongation of the sulcus running parallel to the corpus callosum and can readily be followed in particular in the sagittal section S1a. According to anatomy atlases, when viewed from above it is at the level of the sulcus postcentralis and thus represents the prolongation of the latter in the medial direction. The sinus sagittalis superior is now visibly wider and takes up more and more of the space of the fissura longitudinalis cerebri.

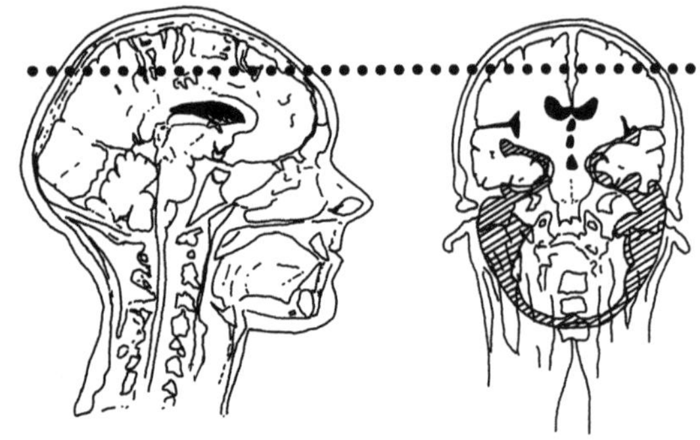

Horizontal section 9 cm above the external acoustic meatus **H 18**

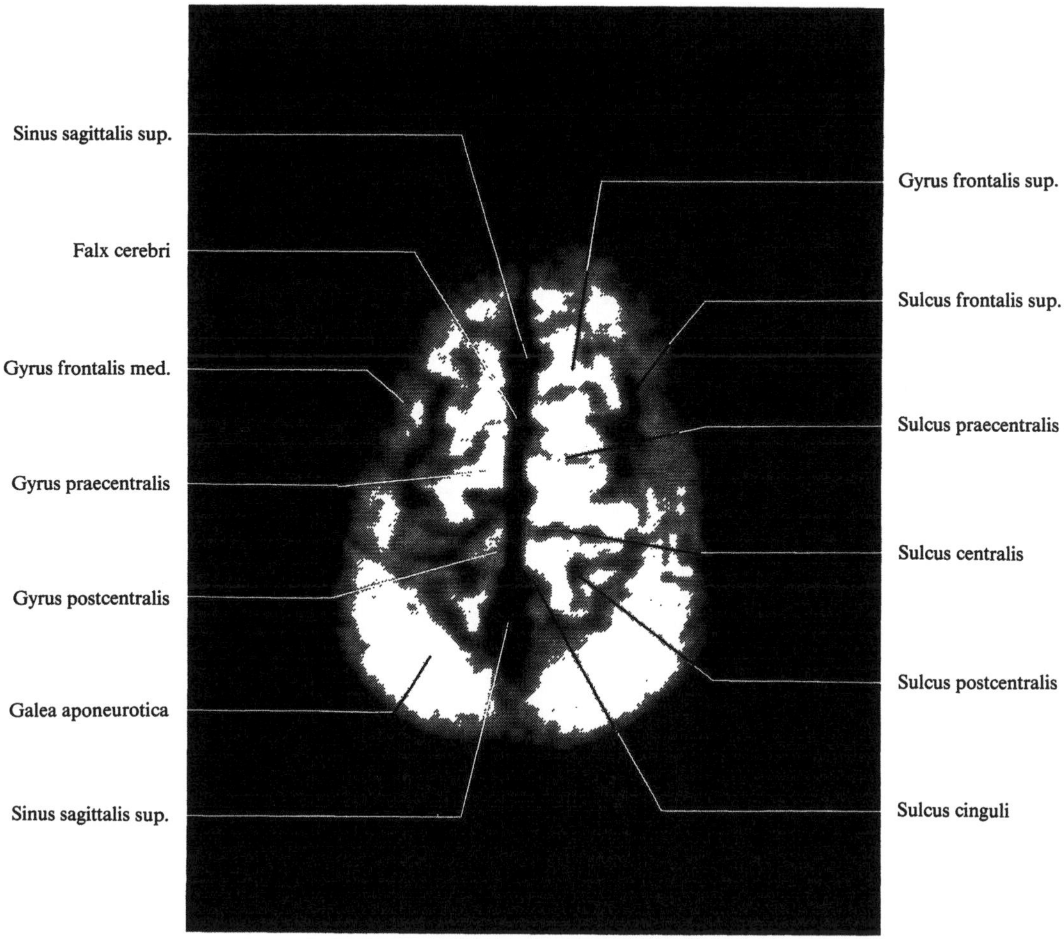

Section H19 shows the most cranial convexity of the cerebral cortex. The convolutions can be followed to the fissura longitudinalis, which is now formed almost exclusively of the sinus sagitallis superior. The gyrus frontalis medius cannot be seen but is located in the blurred frontal margin, while the sulcus frontalis medius can still be readily identified.

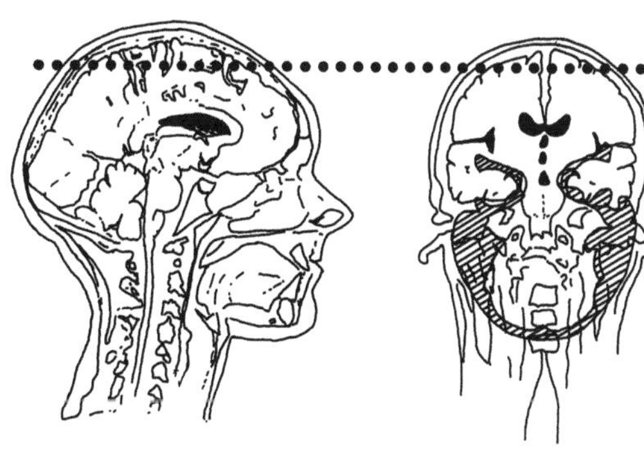

Horizontal section 10 cm above the external acoustic meatus **H 19**

References

Bailes DR, Young IR, Thomas DJ, Straughan K, Bydder GM, Steiner RE (1982) NMR imaging of the brain using spin-echo sequences. Clin Radiol 33: 395–414

Bendel P, Lai CM, Lauterbur PC (1980) 31P-Spectroscopic zeugmatigraphy of phosphorous metabolites. J Magn Reson 38: 343–356

Bloch F (1946) Nuclear induction. Phys Rev 70: 466–474

Bloch F, Hansen WW, Packard M (1946) The nuclear induction experiment. Phys Rev 70: 474–485

Bottomley PA, Andrew ER (1978) RF magnetic field penetration, phase shift and power dissipation in biological tissue: Implications of NMR imaging. Phys Med Biol 23: 630–643

Brant-Zawadzki M, Davis PL, Crooks LE, Mills CM, Norman D, Newton TH, Sheldon P, Kaufman L (1983) NMR demonstration of cerebral abnormalities: Comparison with CT. AJR 140: 847–854

Budinger TF (1981) Nuclear magnetic resonance (NMR) in vivo studies: Known thresholds for health effects. J Comput Assist Tomogr 5/6: 800–811

Bydder GM, Steiner RE (1982) NMR-imaging of the brain. Neuroradiology 23: 231–240

Bydder GM, Steiner RE, Young IR, Hall AS, Thomas DJ, Marshall J, Pallis CA, Legg NJ (1982) Clinical NMR imaging of the brain: 140 cases. AJR 139: 215–236

Crooks LE, Mills CM, Davis PL (1982) Visualisation of cerebral and vascular abnormalities by NMR imaging. The effects of imaging parameters on contrast. Radiology 144: 843–852

Damadian R (1971) Tumor detection by nuclear magnetic resonance. Science 171: 1151–1153

Edelstein WA, Bottomley PA, Hart HR, Smith LS (1983) Signal, noise, and contrast in nuclear magnetic resonance (NMR) imaging. J Comput Assist Tomogr 7/3: 391–401

Evanochko WT, Ng TC, Lilly MB, Lawson AJ, Corbett TH, Durant JR, Glickson JD (1983) In vivo 31-P NMR study of the metabolism of murine mammary 16/C adenocarcinoma and its responce to chemotherapie, X-radiation and hyperthermia. Proc Natl Acad Sci USA 80: 334–338

Gambarelli J, Guerinel G, Chevrot L, Mattei M (1977) Ganzkörper-Computertomographie. Springer, Berlin Heidelberg New York

Habermehl A, Graul EH (1982) Kernspinresonanz-Tomographie. Dtsch Ärztebl 30 C: 17–30

Hahn EL (1950) Spin echos, Phys Rev 80: 580–594

Herfkens R, Davis P, Crooks L, Kaufman L, Price D, Miller T, Margulis AR, Watts J, Hoenninger J, Asakawa M, McRee R (1981) Nuclear magnetic resonance imaging of the abnormal live rat and correlations with tissue characteristics. Radiology 141: 211–218

Holland GN, Hawkes RC, Moore WS (1980) Nuclear magnetic tomography of the brain: Coronal and sagittal sections. J Comput Assist Tomogr 4/4: 429–433

Huk W, Heindel W, Deimling M, Stetter E (1983) Nuclear magnetic resonance (NMR) tomography of the central nervous system: Comparison of two imaging sequences. J Comput Assist Tomogr 7/3: 468–475

Lauterbur PC (1973) Image formation by induced local interaction examples employing nuclear magnetic resonance. Nature 242: 190

New PFJ, Rosen BR, Brady TJ, Buonanno FS, Kistler JP, Burt CT, Hinshaw WS, Newhouse JH, Pohost GM, Taveras JM (1983) Potential hazards and artifacts of ferromagnetic and notferromagnetic surgical and dental materials and devices in nuclear magnetic resonance imaging. Radiology 147: 139–148

Ng TC, Evanochko WT, Hiramoto RN, Ghanta VK, Lilly MB, Lawson AJ, Corbett TH, Durant JR, Glickson JD (1982) 31-P spectroskopy of in vivo tumors. J Magn Reson 49: 271–286

Pavlicek W, Geisinger M, Castle L, Borkowski GP, Meaney TF, Bream BL, Gallagher JH (1983) The effect of nuclear magnetic resonance on patients with cardiac pacemakers. Radiology 147: 149–153

Purcell EM, Torrey HC, Pound RV (1946) Resonance absorption by nuclear magnetic moments in a solid. Phys Rev 69: 37–38

Rupp N, Reiser M, Stetter E (1983) The diagnostic value of morphology on relaxation times in NMR-imaging of the body. Eur J Radiol 3: 68–76

Young IR, Burl M, Clarke GJ, Hall AS, Pasmore T, Collins AG, Smith DT, Orr JS, Bydder GM, Doyle FH, Greenspan RH, Steiner RE (1981) Magnetic resonance properties of hydrogen: Imaging the posterior fossa. AJR 137: 895–901

Young IR, Bailes DR, Burl M, Collins AG, Smith DT, McDonnell MJ, Orr JS, Banks LM, Bydder GM, Greenspan RH, Steiner RE (1982) Initial clinical evaluation of a whole body nuclear magnetic resonance (NMR) tomograph. J Comput Assist Tomogr 6/1: 1–18

Young IR, Randell CP, Kaplan PW, James A, Bydder GM, Steiner RE (1983) Nuclear magnetic resonance (NMR) imaging in white matter disease of the brain using spin-echo sequences. J Comput Assist Tomogr 7/2: 290–294

Subject Index

Numbers in *italics* refer to the coordinates in the S, F, H-system (see Chap. 3).

MIX
Papier aus verantwortungsvollen Quellen
Paper from responsible sources
FSC® C105338

If you have any concerns about our products,
you can contact us on
ProductSafety@springernature.com

In case Publisher is established outside the EU,
the EU authorized representative is:
Springer Nature Customer Service Center GmbH
Europaplatz 3, 69115 Heidelberg, Germany

Printed by Libri Plureos GmbH
in Hamburg, Germany